London Mathematical Society Lecture Note Series 3

Convex Polytopes and the Upper Bound Conjecture

P. McMULLEN and G.C. SHEPHARD
in collaboration with
J. E. REEVE and A.A. BALL

CAMBRIDGE AT THE UNIVERSITY PRESS 1971

Published by The Syndics of the Cambridge University Press
Bentley House, 200 Euston Road, London N.W. 1
American Branch: 32 East 57th Street, New York, N.Y. 10022

© Cambridge University Press 1971

Library of Congress Catalogue Card No.: 75-130909

ISBN: 0 521 08017 7

Printed offset in Great Britain by
Alden & Mowbray Ltd at the Alden Press, Oxford

Contents

		Page
Preface		iii

Chapter 1	BASIC PROPERTIES OF CONVEX SETS	
1.1	Linear, Affine and Convex Dependence	1
1.2	Transformations	14
1.3	Three Basic Theorems	22
1.4	Support Properties	28

Chapter 2	CONVEX POLYTOPES	
2.1	The Faces of a Convex Polytope	39
2.2	Polarity and Duality	61
2.3	Some Special Types of Polytopes	74
	(i) Simplices	74
	(ii) Pyramids	75
	(iii) Bipyramids	77
	(iv) Prisms	78
	(v) Simplicial and Simple Polytopes	81
	(vi) Cyclic polytopes	82
	(vii) Neighbourly Polytopes	90
2.4	Euler's Theorem and the Dehn-Sommerville Equations	93
2.5	Pulling the Vertices of a Polytope	112

Chapter 3		GALE DIAGRAMS, AND POLYTOPES WITH FEW VERTICES	119
	3.1	Gale Transforms: Geometric Formulation	120
	3.2	Gale Transforms: Algebraic Formulation	127
	3.3	Gale Diagrams	134
	3.4	Polytopes with $d+2$ and $d+3$ Vertices	143
Chapter 4		THE UPPER BOUND CONJECTURE FOR SPHERICAL COMPLEXES	
	4.1	Spherical Complexes	152
	4.2	Particular Cases of the U.B.C.	154
Chapter 5		THE UPPER BOUND CONJECTURE FOR POLYTOPES	
	5.1	Reformulations of the Dehn-Sommerville Equations	169
	5.2	Shelling the Boundary Complex	173
	5.3	Remarks	177
References			180
Index			182

Preface

Polytopes are the analogues, in d-dimensional euclidean space E^d ($d \geq 0$) of the familiar polygons in E^2 and polyhedra in E^3. They were discovered (for $d \geq 4$) just over a century ago and many famous mathematicians have made contributions to the subject. Much of the early work was concerned with the metrical properties of polytopes, and it is only during the last few years that attention has been focussed on problems of a combinatorial nature. One such problem is that of proving the Upper Bound Conjecture.

This conjecture is concerned with the number of faces (of various dimensions) that a polytope may possess. It states that the maximum possible number of j-dimensional faces of a d-dimensional polytope P with v vertices is attained when P is a cyclic polytope, a special type of polytope which will be described in §2.3 (vi). At the time these notes were written quite a large number of special cases of the conjecture had been established. Many of the methods used were applicable to spherical complexes as well as to polytopes, and a brief account of these will be given in Chapter 4. However, in July 1970, while these notes were in press, one of the authors (P. McMullen) succeeded in proving the conjecture is true generally for polytopes, and this proof is presented in Chapter 5. It depends upon the interesting process, recently described by Bruggesser and Mani, known as 'shelling' a polytope.

These lecture notes are a record of a research seminar held at the University of East Anglia, Norwich, during the early part of 1968. The aim of the seminar was to provide an introduction to the theory of convex polytopes. Originally it was intended to follow closely the relevant parts of the book Convex Polytopes by Branko Grünbaum (J. Wiley and Sons, Ltd., London-New York-Sydney, 1967) but it soon became clear that there were going to be considerable divergences. For example, our treatment of the support properties of convex sets using the nearest-point map* (§1.3), our solution of the Dehn-Sommerville equations based on the method of contour integration devised by Ian MacDonald (§2.4), and our treatment of polytopes with few vertices (§3.4), differ greatly from those given in Grünbaum's book. For this reason it seemed worth while to compile these notes and publish them. We would like, however, to record our great indebtedness to Branko Grünbaum, not only on account of the book mentioned above, but also for his encouragement and help.

In conclusion, it should be explained that the seminar lectures were given by the four persons named on the title page. These notes, which represent a considerable expansion of the material covered in the lectures, were written by the two authors, who are responsible for the presentation, as well as the accuracy of the material.

P. M. and G. C. S.

---

* We are indebted to Professor G. Ewald for drawing our attention to this approach.

1. Basic Properties of Convex Sets

1.1 LINEAR, AFFINE AND CONVEX DEPENDENCE

Throughout we shall work in d-dimensional euclidean space E^d, writing

$$x = (\xi_1, \ldots, \xi_d), \quad (\xi_i \in R, \; i = 1, \ldots, d),$$

for a point of the space. It is convenient in this subject to use d for the dimension, rather than n which is useful for other purposes. E^d is, of course, an inner-product space, and we denote the inner product of the points, or vectors, x and $y = (\eta_1, \ldots, \eta_d)$ by

$$\langle x, y \rangle = \sum_{i=1}^{d} \xi_i \eta_i .$$

The <u>norm</u>, or <u>distance</u>, between the points x and y is denoted by

$$\|x - y\| = \langle x - y, x - y \rangle^{\frac{1}{2}} .$$

We recall that x is <u>linearly dependent</u> on the subset X of E^d if there are points $x_1, \ldots, x_r$ of X and scalars (real numbers) $\lambda_1, \ldots, \lambda_r$ such that

$$x = \lambda_1 x_1 + \ldots + \lambda_r x_r .$$

If we impose the additional condition

$$\lambda_1 + \ldots + \lambda_r = 1,$$

then x is <u>affinely dependent</u> on X. If also

$$\lambda_i \geq 0, \ i = 1, \ldots, r,$$

then x is said to be <u>convexly dependent</u> on the set X. Thus linear, affine and convex dependence are increasingly strong conditions on the type of linear combination that can be taken to represent x.

Just as we define linearly dependent and independent sets of points, so also we define affinely dependent and independent sets. We say that $X \subseteq E^d$ is <u>affinely dependent</u> if there exists a relation of the form

$$\left. \begin{array}{l} \lambda_1 x_1 + \ldots + \lambda_r x_r = o, \\ \lambda_1 + \ldots + \lambda_r = 0, \end{array} \right\}$$

for some $x_1, \ldots, x_r \in X$ and $\lambda_1, \ldots, \lambda_r \in R$, with at least one λ_i non-zero. (We write o for the zero vector: $o = (0, \ldots, 0)$.) If no such relation exists, then X is said to be <u>affinely independent</u>. A useful criterion for affine independence is the following: if $x_i = (\xi_{i1}, \ldots, \xi_{id})$, then $\{x_1, \ldots, x_r\}$ is an affinely independent set of points if and only if the matrix

$$\begin{pmatrix} 1 & \xi_{11} & \cdots & \xi_{1d} \\ 1 & \xi_{21} & \cdots & \xi_{2d} \\ \cdot & \cdot & & \cdot \\ \cdot & \cdot & & \cdot \\ \cdot & \cdot & & \cdot \\ 1 & \xi_{r1} & \cdots & \xi_{rd} \end{pmatrix}$$

has rank r. This follows immediately from the definitions, and, in particular, since every subset of an affinely independent set is affinely independent, implies that every maximal affinely independent subset of E^d contains exactly $d + 1$ points.

On the other hand, there is no sensible way of defining a set of points to be convexly dependent or independent. Consequently, whereas there is a great deal of similarity between the linear and affine theories, the results of convexity are very different.

A (linear) subspace of E^d is any subset L of E^d that is closed under the operation of taking linear combinations of (finite) sets of points of L. An affine subspace of E^d is any subset which is closed under the operation of taking affine combinations (that is, linear combinations whose coefficients sum to 1). A convex set in E^d is any subset which is closed under the operation of taking convex combinations (that is, affine combinations with non-negative coefficients). We see immediately that every linear subspace is an affine subspace, and every affine subspace is a convex set.

For example, if we think of E^3 geometrically, then proper linear subspaces are represented by lines or planes through the origin o. The proper affine subspaces are represented by points, lines or planes in any position. Our intuitive idea of a convex set is formalized by the following theorem.

Theorem 1. *A subset K of E^d is convex if and only if for all $x_0, x_1 \in K$, and $0 \leq \lambda \leq 1$, the point $x = (1 - \lambda) x_0 + \lambda x_1$ also belongs to K.*

In other words, a convex set is completely characterized by the fact that it is closed under the operation of taking convex combinations of sets of two points (instead of arbitrarily large finite subsets, as in the definition). Geometrically, it means that K is convex if and only if it contains all the line segments whose endpoints lie in K. This, in turn, has the trivial implication that a convex set is (path-) connected.

The proof of the theorem is simple. Clearly the given condition is necessary. To show that it is sufficient, consider a point x which is a convex combination of points of K, say

$$x = \lambda_1 x_1 + \ldots + \lambda_r x_r$$

where

$$x_i \in K, \quad \lambda_i \geq 0 \ (i = 1, \ldots, r), \quad \sum_{i=1}^{r} \lambda_i = 1.$$

Assume without loss of generality that $\lambda_i \neq 0$ $(i = 1, \ldots, r)$; otherwise delete the corresponding terms. If the condition of the theorem holds, then

$$y_1 = \frac{\lambda_1}{\lambda_1 + \lambda_2} x_1 + \frac{\lambda_2}{\lambda_1 + \lambda_2} x_2 \in K,$$

so that

$$y_2 = \frac{\lambda_1 + \lambda_2}{\lambda_1 + \lambda_2 + \lambda_3} y_1 + \frac{\lambda_3}{\lambda_1 + \lambda_2 + \lambda_3} x_3 \in K,$$

and so on. By induction,

$$x = \frac{\lambda_1 + \ldots + \lambda_{r-1}}{\lambda_1 + \ldots + \lambda_r} y_{r-2} + \frac{\lambda_r}{\lambda_1 + \ldots + \lambda_r} x_r \in K.$$

We have thus shown that an arbitrary convex combination of points of K belongs to K, so that, by definition, the set K is convex. This proves the theorem.

An example of a convex set in E^d to which we shall need to refer is the closed d-ball $B(a, \rho)$ with centre a and radius ρ ($\rho \geq 0$), defined by

$$B(a, \rho) = \{x \in E^d \mid \|x - a\| \leq \rho\}.$$

The fact that $B(a, \rho)$ is convex follows at once from Theorem 1 and the triangle inequality; for if $x_0, x_1 \in B(a, \rho)$ and $0 \leq \lambda \leq 1$, then

$$\|((1 - \lambda)x_0 + \lambda x_1) - a\| \leq (1 - \lambda)\|x_0 - a\| + \lambda\|x_1 - a\|$$
$$\leq (1 - \lambda)\rho + \lambda\rho = \rho.$$

The interior of $B(a, \rho)$ ($\rho > 0$), namely

$$\{x \in E^d \mid \|x - a\| < \rho\},$$

is also convex; this is called an open d-ball, or the ρ-neighbourhood of a. The boundary of $B(a, \rho)$ is the (d - 1)-sphere $S(a, \rho)$, so that

$$S(a, \rho) = \{x \in E^d \mid \|x - a\| = \rho\}.$$

Note that (for $\rho > 0$) $S(a, \rho)$ is not a convex set.

Notice that the empty set $\emptyset$ is convex, for in this case the condition of Theorem 1 is satisfied vacuously.

It is well-known that in the theory of vector spaces there are basis theorems which lead to the definition of (linear) dimension. There are exact analogues for affine spaces, and we now quote these without proof.

Let X be any given set of points (finite or infinite) in E^d. It is clear that the intersection of affine subspaces is again an affine subspace; the intersection of the affine subspaces of E^d containing X is, in an obvious intuitive sense, the 'smallest' affine subspace containing X. It is called the <u>affine hull</u> of X, and is denoted by aff X. In fact, aff X is also the set of all affine combinations of points of X. If X is a subset of an affine subspace L of E^d such that L = aff X, then we say that X <u>affinely spans</u> L; it can be shown that every affine subspace L can be written in the form L = aff X for some <u>finite</u> set X. If X is a minimal subset affinely spanning L, then it is also affinely independent, and X is then called an <u>affine basis</u> of L. Every affine basis of a given affine subspace L contains the same number of points; if we denote this number by n + 1, then n is called the <u>affine dimension</u> of L, and we write dim L = n. Notice that this leads to the useful convention dim $\emptyset$ = -1 for the empty set $\emptyset$.

The <u>linear hull</u> of X, denoted by lin X, is defined in a similar manner. We have already remarked that every linear subspace of E^d is also an affine subspace. If L has (linear) dimension n, and so some (linear) basis $\{x_1, \ldots, x_n\}$ consisting of n points, then $\{o, x_1, \ldots, x_n\}$ is an affine basis of L (as an affine subspace), and so the affine dimension of L is also n. Hence the use of the same notation dim L for both the linear and affine dimensions of L leads to no ambiguities.

We shall also need the following definitions. A point $x \in E^d$ is said to be <u>positively dependent</u> on the subset $X \subseteq E^d$

if $x = \lambda_1 x_1 + \ldots + \lambda_r x_r$, for some $x_1, \ldots, x_r \in X$ and some real numbers $\lambda_1, \ldots, \lambda_r \geq 0$. A subset of E^d closed under the operations of taking such non-negative linear combinations is called a <u>convex cone</u>. It is clear that a linear subspace of E^d is a convex cone.

The set of all non-negative linear combinations of points of a subset X of E^d is a convex cone, called the <u>positive hull</u> of X, and denoted by pos X. Let L be a linear subspace of E^d. If the subset X of E^d is such that $L = $ pos X, then we say that X <u>positively spans</u> L.

Following the usual terminology, we shall call an affine subspace of zero dimensions a <u>point</u>, since it consists of a single vector $\{x\}$. (Strictly we should call it a <u>point-set</u>, but it is not usually worth while to distinguish between a point and the point-set to which it belongs.) An affine subspace of 1 dimension is called a <u>line</u>, and, in E^d, an affine subspace of $d - 1$ dimensions is called a <u>hyperplane</u>. A hyperplane will usually be denoted by the letter H, with or without subscripts, and it can be conveniently written in the form

$$H = \{x \in E^d \mid \langle x, a \rangle = \gamma\},$$

where a is a non-zero vector, and $\gamma \in R$. We say that a is <u>normal</u> to H, and define two hyperplanes to be <u>parallel</u> if their normals are scalar multiples of one another. Parallel hyperplanes either coincide or have empty intersection. (We shall give an alternative, and more general definition of parallelism in the next section.)

The intersection theory of affine subspaces is slightly more complicated than that of linear subspaces, due to the fact that, if L_1 and L_2 are affine subspaces, then, as we have just seen,

$L_1 \cap L_2 = \emptyset$ is possible. If $L_1 \cap L_2 \neq \emptyset$, then we may choose our coordinate system so that $o \in L_1 \cap L_2$, so that L_1 and L_2 become linear subspaces. Hence:

Theorem 2. <u>If L_1 and L_2 are affine subspaces of E^d, then $L_1 \cap L_2$ is an affine subspace, and either</u>

(i) $L_1 \cap L_2 = \emptyset$, <u>or</u>

(ii) $\dim(L_1 \cap L_2) \geq \dim L_1 + \dim L_2 - d$.

The <u>dimension</u> of a convex set K, also denoted by dim K, is defined to be the dimension of the subspace aff K, or equivalently to be one less than the maximum number of affinely independent points in K. The analogy between convex sets and linear and affine subspaces can be carried a little further, as the next two theorems show.

Theorem 3. <u>Let $\{K_i | i \in I\}$ be any non-empty family of convex sets in E^d. Then $\bigcap_{i \in I} K_i$ is convex.</u>

Proof. Let $x_0, x_1 \in K_i$ (for all $i \in I$), and let $0 \leq \lambda \leq 1$. Since each K_i is convex, $(1-\lambda)x_0 + \lambda x_1 \in K_i$ (all i). Thus $x_0 \in \bigcap_i K_i$, $x_1 \in \bigcap_i K_i$ and $0 \leq \lambda \leq 1$ implies that $(1-\lambda)x_0 + \lambda x_1 \in \bigcap_i K_i$, and so, by Theorem 1, $\bigcap_i K_i$ is convex.

Let X be any subset of E^d, and let $\{K_i\}$ be the family of all convex sets which contain X. This family is non-empty, because $E^d \supseteq X$ is convex. Then $\bigcap_i K_i$ is called the <u>convex hull</u> of X, and is denoted by conv X. It is, in an obvious sense, the 'smallest' convex set containing X.

Corresponding to the result about spanning sets of affine subspaces, we have the following:

Theorem 4. <u>The convex hull of a set X is the set of all convex combinations of (finite) subsets of X. That is,</u>

$$\operatorname{conv} X = \{x \in E^d \mid x = \lambda_1 x_1 + \ldots + \lambda_r x_r,$$
$$1 \le r < \infty, \ x_i \in X, \ \lambda_i \ge 0, \ \sum_{i=1}^{r} \lambda_i = 1\}.$$

Proof. Denote the set on the right side of the above equation by $M(X)$. Then, if

$$y_0 = \mu_1 x_1 + \ldots + \mu_r x_r, \ x_i \in X, \ \mu_i \ge 0, \ \sum \mu_i = 1,$$
$$y_1 = \nu_1 x_1 + \ldots + \nu_r x_r, \ \nu_i \ge 0, \ \sum \nu_i = 1,$$

are arbitrary elements of $M(X)$ (where zero coefficients μ_i and ν_i are inserted if necessary, so that $x_1, \ldots, x_r$ all occur in the expressions for both y_0 and y_1), then for $0 \le \lambda \le 1$,

$$y = (1 - \lambda)y_0 + \lambda y_1 = \sum_{i=1}^{r} ((1 - \lambda)\mu_i + \lambda \nu_i) x_i$$

is also a convex combination of points of X; that is, $y \in M(X)$. Hence by Theorem 1, $M(X)$ is convex.

Since $X \subseteq M(X)$, it follows from the definition that $M(X) \supseteq \operatorname{conv} X$. On the other hand, if $x = \lambda_1 x_1 + \ldots + \lambda_r x_r$ ($x_i \in X$, $\lambda_i \ge 0$, $\sum \lambda_i = 1$) is any point of $M(X)$, then any convex set containing X contains $x_1, \ldots, x_r$, and so also contains x. Hence $\operatorname{conv} X \supseteq M(X)$. Thus $\operatorname{conv} X = M(X)$ and the theorem is proved.

However, at this point the analogy between linear and affine subspaces and convex sets breaks down, for the d-ball $B(a, \rho)$ is an example of a closed bounded convex set which is, nevertheless, not the convex hull of a finite (or even denumerable) set of points. Bearing this in mind, we make the following definition:

Definition. The convex hull of a finite set of points is called a <u>convex polytope</u>.

It is easy to see that a convex polytope is closed and bounded (see also the corollary to Carathéodory's Theorem 11). In the sequel we shall be almost entirely concerned with closed bounded sets; unbounded sets will only be mentioned briefly in the next chapter. For brevity we shall usually use the word 'polytope' instead of 'convex polytope'; no confusion can arise, since no other type of polytope will be considered.

A polytope P is called an r-<u>polytope</u> if $\dim P = r$. If $X = \{x_0, \ldots, x_r\}$ is a set of affinely independent points in E^d, then $T^r = \text{conv } X$ is a particular type of r-polytope known as an r-<u>simplex</u>. The properties of simplices will be discussed in §2.3. If we write

$$x = \frac{1}{r+1} (x_0 + \ldots + x_r),$$

then x is called the <u>centroid</u> of T^r, and by elementary algebra it is easy to show that there exists an $\eta > 0$ such that

$$B(x, \eta) \cap \text{aff } T^r \subseteq T^r.$$

Hence the interior of T^r, relative to aff T^r, is non-empty. More generally:

Theorem 5. *An r-dimensional convex set K $(r \geq 0)$ has non-empty interior relative to aff K.*

Proof. Since K is r-dimensional, we can find $r+1$ affinely independent points $x_0, \ldots, x_r \in K$. Then $T^r = \text{conv}\{x_0, \ldots, x_r\}$ has non-empty interior relative to aff $T^r =$ aff K, and since $K \supseteq T^r$, we see that K has this property also.

It will be convenient to use the notations int K for the interior of K, and relint K for the relative interior of K, relative, that is, to aff K. Thus Theorem 5 can be simply stated: For every non-empty convex set K, relint $K \neq \emptyset$.

We shall need certain results about the intersections of lines with convex sets. Let K be a closed bounded convex set (dim $K > 0$), and let $x \in$ relint K. Since K contains some sphere $B(x, \rho) \cap$ aff K, it is clear that any line L in aff K through x must meet K in an interval with x in its relative interior. In other words, letting relbd K denote the relative boundary of K (relative, that is, to aff K), we see that $L \cap$ relbd K consists of two points.

It follows at once from this that if $x \in$ relint K and $y \in$ aff $K \setminus K$, then conv $\{x, y\} \cap$ relbd K consists of one point, and this point lies in relint conv $\{x, y\}$. Also aff $\{x, y\} \cap$ relbd K consists of two points, one of which is that just mentioned, and the other belongs to aff $\{x, y\} \setminus$ conv $\{x, y\}$.

In the next section, we shall require the following theorem.

Theorem 6. *Let $X = \{x_1, \ldots, x_r\}$ be a finite set of points in E^d, and let P be the convex polytope*

$$P = \text{conv } X.$$

Then $x \in \operatorname{relint} P$ if and only if x can be written in the form

$$x = \sum_{i=1}^{r} \lambda_i x_i, \quad \lambda_i > 0, \quad \sum_{i=1}^{r} \lambda_i = 1,$$

that is, a convex combination of X with strictly positive coefficients.

Proof. We first show the condition is sufficient. Let x be such a positive convex combination, and suppose, without loss of generality, that $\{x_1, \ldots, x_k\}$ is a maximal affinely independent subset of X (dim $P = k - 1$). If

$$\lambda = \lambda_1 + \ldots + \lambda_k,$$

then

$$y_1 = \frac{1}{\lambda} \sum_{i=1}^{k} \lambda_i x_i$$

is clearly a relatively interior point of the simplex conv $\{x_1, \ldots, x_k\}$. (The points of relbd conv $\{x_1, \ldots, x_k\}$ are precisely those points that can be written as a convex combination of $x_1, \ldots, x_k$ with at least one zero coefficient.) Since conv $\{x_1, \ldots, x_k\} \subseteq P$, it follows that $y_1 \in \operatorname{relint} P$ also. If $k = r$, there is no more to prove; otherwise let

$$y_0 = \frac{1}{1-\lambda} \sum_{i=k+1}^{r} \lambda_i x_i \in P,$$

and so

$$x = (1 - \lambda)y_0 + \lambda y_1 \quad (0 < \lambda < 1)$$

belongs to relint P also.

The condition is also necessary. Let $x \in$ relint P, and let y_0 be the point.

$$y_0 = \frac{1}{r} \sum_{i=1}^{r} x_i ,$$

then $y_0 \in P$ (actually, as we have just seen, to relint P). If $x = y_0$, there is no more to prove. Otherwise, let L be the line

$$L = \text{aff } \{x, y_0\}$$

since $x \in$ relint P, there is a further point $y_1 \in P$,

$$y_1 = \sum_{i=1}^{r} \mu_i x_i , \quad \mu_i \geq 0 , \quad \sum_{i=1}^{r} \mu_i = 1 ,$$

such that $x \in$ relint conv $\{y_0, y_1\}$. (See Figure 1.)

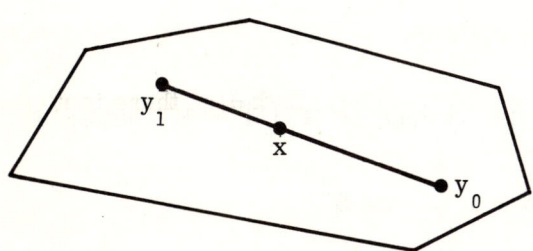

Figure 1

That is, for some $0 < \nu < 1$,

$$x = (1 - \nu)y_0 + \nu y_1,$$

so that

$$x = \sum_{i=1}^{r} \lambda_i x_i,$$

where

$$\lambda_i = (1 - \nu)\frac{1}{r} + \nu\mu_i > 0 \quad (i = 1, \ldots, r).$$

which proves the theorem.

1.2   TRANSFORMATIONS

In establishing the properties of convex polytopes, we use various types of transformation, and this section is devoted to a brief discussion of these.

Definition. A mapping $\Gamma : E^d \to E^k$ defined by

$$x\Gamma = x\Lambda + a \quad (x \in E^d),$$

where $\Lambda : E^d \to E^k$ is a linear transformation, and $a \in E^k$ is a fixed vector, is called an <u>affine transformation</u>. If Λ is singular (non-singular), then the affine transformation Γ is called <u>singular</u> (<u>non-singular</u>, respectively). A non-singular affine transformation is sometimes called an <u>affinity</u>. If Λ is the identity linear transformation, then Γ is called a <u>translation</u>, and if Λ is an orthogonal transformation, then Γ is called a <u>congruent transformation</u>.

Affine transformations map affinely dependent sets of points into affinely dependent sets, and so map affine subspaces into affine subspaces. An affinity also maps affinely independent sets into affinely independent sets, and so maps each affine subspace into a subspace of the same dimension. Congruent transformations are those affine transformations Γ which preserve length; that is, for all $x, y \in E^d$,

$$\|x\Gamma - y\Gamma\| = \|x - y\|.$$

Such transformations are necessarily non-singular.

One particular congruent transformation of importance is the <u>central reflection</u> of E^d, defined by

$$x \to -x \quad (x \in E^d).$$

Two affine subspaces L_1 and L_2 are said to be <u>parallel</u> if there is a translation Γ such that $L_1 \Gamma = L_2$. Clearly parallel subspaces have the same dimension. This definition agrees with that of §1.1 for hyperplanes: if H_1 and H_2 are parallel hyperplanes, and $x_1 \in H_1$, $x_2 \in H_2$, then the translation Γ, defined by

$$x\Gamma = x + (x_2 - x_1),$$

is such that $H_1 \Gamma = H_2$.

From our point of view, one of the most important properties of affine transformations is that they preserve convexity. Precisely, for any convex set $K \subseteq E^d$, and any affine transformation Γ of E^d,

$$K\Gamma = \{x\Gamma \mid x \in K\}$$

is also a convex set. This follows immediately from the definitions. In particular, if $K = \text{conv } X$, then $K\Gamma = \text{conv}(X\Gamma)$, and so the image of a convex polytope under an affine transformation is also a convex polytope. If X is any set, and Γ any affinity, then $X\Gamma$ is said to be affinely equivalent to X.

Klein, in his famous Erlanger Program, defined a geometry to be the study of those properties of a space which are invariant under a given group of transformations of that space. Euclidean geometry is concerned with properties that are invariant under congruent transformation, and affine geometry with those that are invariant under affinities. Hence, according to Klein's definition, convexity belongs to affine geometry (as well as to euclidean geometry). However, the reader will notice that we shall often use metrical methods to obtain results of an affine nature. One example of this occurs in §1.4, where we shall establish the support properties of convex sets, which properly belong to affine geometry, using a metrical tool, namely the nearest point map.

An analogous situation will occur in §2.4. Here we shall prove Euler's Theorem, which is a topological property of convex polytopes, using the methods of affine geometry.

In analogy to the corresponding result for linear transformations, we have the following:

Theorem 7. <u>Let $\{x_0, \ldots, x_d\}$ be an affinely independent set of points (affine basis) of E^d, and $\{y_0, \ldots, y_d\}$ any set of points of E^k. Then there is an affine transformation Γ of E^d such that</u>

$$x_i \Gamma = y_i, \quad i = 0, \ldots, d.$$

If, in addition, $\{y_0, \ldots, y_d\}$ is also affinely independent, then Γ is non-singular and unique.

Proof. From the definition of affine independence, we see that the set

$$\{x_1 - x_0, \ldots, x_d - x_0\}$$

is linearly independent. From vector space theory, there is a linear transformation Λ such that

$$(x_i - x_0)\Lambda = y_i - y_0, \quad i = 1, \ldots, d.$$

Then the affine transformation Γ, defined by

$$x\Gamma = x\Lambda + (y_0 - x_0\Lambda)$$

is seen to have the required properties. The remaining assertions of the theorem follow directly from the corresponding theorems of vector space theory.

An immediate consequence of Theorem 7 is that if T_1^r and T_2^r are two r-simplices in E^d ($d \geq r$), then there exists an affine transformation Γ of E^d such that $T_1^r \Gamma = T_2^r$. Γ will be non-singular if $r = d$, and can be chosen to be non-singular if $r < d$. If $r = d$, then Γ will be determined uniquely if we specify which of the $d+1$ points used in defining T_2^d is the image of each of the $d+1$ points used in defining T_1^d. Summarizing:

Corollary. <u>Any two r-simplices in E^d ($r \leq d$) are affinely equivalent.</u>

For the most part we shall only consider affinities, but one type of singular affine transformation is of great importance. Two affine subspaces L^r and L^s of E^d are <u>perpendicular</u> if $L^r \cap L^s \neq \emptyset$, and, if we choose the coordinate system in E^d so that $o \in L^r \cap L^s$, then $\langle y, z \rangle = 0$ for all $y \in L^r$ and $z \in L^s$. This implies that $r + s \leq d$. If L^r and L^s are complementary (that is, $r + s = d$), then every $x \in E^d$ can be written (uniquely) in the form $x = y + z$, with $y \in L^r$ and $z \in L^s$. The mapping $\Pi : E^d \to L^r$ defined by $x\Pi = y$ is called the <u>orthogonal projection</u> onto L^r; it is, for $r < d$, a singular affine transformation. An orthogonal projection Π, in addition to having all the usual properties of affine transformations, tends to reduce lengths; that is, for all $x, y \in E^d$,

$$\|x\Pi - y\Pi\| \leq \|x - y\|.$$

In one application we shall also need the concept of a <u>projective transformation</u>. This is a mapping $\Theta : E^d \to E^k$ of the form

$$x\Theta = \frac{x\Gamma}{\langle x, a \rangle + \beta},$$

where $\Gamma : E^d \to E^k$ is an affine transformation, $a \in E^d$ is a fixed vector, and β is a scalar, such that $(a, \beta) \neq o$ (where, if $a = (\alpha_1, \ldots, \alpha_d)$, we write (a, β) for $(\alpha_1, \ldots, \alpha_d, \beta)$). If $a = o$, then Θ is an affine transformation; if $a \neq o$, the domain of the transformation Θ is $E^d \setminus H$, where H is the hyperplane

$$H = \{x \in E^d \mid \langle x, a \rangle + \beta = 0\}.$$

In the usual terminology, we say that H is mapped into the 'hyperplane at infinity', that is, the hyperplane that must be adjoined to E^d to make it into a projective space. Θ is non-singular if the corresponding linear transformation $\overline{\Theta}: E^{d+1} \to E^{k+1}$ defined by

$$(x, 1)\overline{\Theta} = (x\Gamma, \langle x, a \rangle + \beta) \quad (x \in E^d),$$

is non-singular (one-to-one); otherwise Θ is singular.

Let $X = \{x_1, \ldots, x_r\}$ be an affinely dependent set of points in E^d, such that $H \cap X = \emptyset$. That is, for some $\lambda_1, \ldots, \lambda_r$, not all zero,

$$\sum_{i=1}^{r} \lambda_i x_i = 0, \quad \sum_{i=1}^{r} \lambda_i = 0.$$

Then

$$x_i \Theta = \frac{x_i \Gamma}{\langle x_i, a \rangle + \beta}, \quad (i = 1, \ldots, r),$$

so that

$$\sum_{i=1}^{r} \lambda_i (\langle x_i, a \rangle + \beta)(x_i \Theta) = \sum_{i=1}^{r} \lambda_i (x_i \Gamma)$$

$$= \sum_{i=1}^{r} (\lambda_i x_i)\Gamma = 0$$

since Γ is an affine transformation. We also have

$$\sum_{i=1}^{r} \lambda_i (\langle x_i, a \rangle + \beta) = \langle \sum_{i=1}^{r} \lambda_i x_i, a \rangle + \beta \sum_{i=1}^{r} \lambda_i$$
$$= 0.$$

It follows that $X\Theta$ is also affinely dependent. In other words (ignoring the awkward fact that some points are taken to, and come from, infinity), <u>the projective image of an affine subspace is an affine subspace.</u>

A projective transformation Θ is said to be <u>permissible</u> for a set X if $H \cap X = \emptyset$. It follows immediately from the order properties of points on a real projective line that if Θ is a projective transformation permissible for a convex set K, then $K\Theta$ is also convex. The main property of projective transformations we shall require is the following:

Theorem 8. <u>Let $\{x_0, \ldots, x_d\}$ and $\{y_0, \ldots, y_d\}$ be two affinely independent sets of points in E^d, and let</u>

$$x' \in \text{int conv } \{x_0, \ldots, x_d\},$$

$$y' \in \text{int conv } \{y_0, \ldots, y_d\}.$$

<u>Then there exists a projective transformation Θ of E^d, permissible for</u> $\text{conv } \{x_0, \ldots, x_d\}$, <u>such that</u> $x_i \Theta = y_i$ $(i = 0, \ldots, d)$, <u>and</u> $x'\Theta = y'$.

Proof. By Theorem 6, we can write $x' = \sum_{i=0}^{d} \lambda_i x_i$, $\lambda_i > 0$, $\sum_{i=0}^{d} \lambda_i = 1$, and $y' = \sum_{i=0}^{d} \mu_i y_i$, $\mu_i > 0$, $\sum_{i=0}^{d} \mu_i = 1$. By Theorem 7, since $\{x_0, \ldots, x_d\}$ is affinely independent, there is an affine transformation Γ such that

$$x_i \Gamma = \frac{\mu_i}{\lambda_i} y_i, \quad i = 0, \ldots, d.$$

(Note that $\{\frac{\mu_0}{\lambda_0} y_0, \ldots, \frac{\mu_d}{\lambda_d} y_d\}$ need not be affinely independent, so Γ can be singular.) Let $z_0, \ldots, z_d$ be the vectors of E^{d+1}

$$z_i = \frac{\lambda_i}{\mu_i}(x_i, 1), \quad i = 0, \ldots, d,$$

where, as before, if $x = (\xi_1, \ldots, \xi_d)$ we write $(x, 1)$ to mean $(\xi_1, \ldots, \xi_d, 1)$. Then since $\{x_0, \ldots, x_d\}$ is affinely independent, it follows from the definitions that $\{z_0, \ldots, z_d\}$ is linearly independent. Hence the vectors $z_0, \ldots, z_d$ span a hyperplane H' in E^{d+1}, which, since $o \notin H'$, can be written:

$$H' = \{z \in E^{d+1} \mid \langle z, c \rangle = 1\}.$$

Let $c = (b, \gamma)$, with $b \in E^d$. Then for $i = 0, \ldots, d$,

$$\langle x_i, b \rangle + \gamma = \frac{\mu_i}{\lambda_i} \langle z_i, c \rangle = \frac{\mu_i}{\lambda_i},$$

and so

$$\langle x', b \rangle + \gamma = \sum_{i=0}^{d} \lambda_i \frac{\mu_i}{\lambda_i} = \sum_{i=0}^{d} \mu_i = 1.$$

Clearly $\langle x, b \rangle + \gamma > 0$ for each $x \in \text{conv}\{x_0, \ldots, x_d\}$; thus if

$$H = \{x \in E^d \mid \langle x, b \rangle + \gamma = 0\},$$

then $H \cap \text{conv}\{x_0,\ldots,x_d\} = \emptyset$. Hence the projective transformation Θ:

$$x\Theta = \frac{x\Gamma}{\langle x, b \rangle + \gamma}$$

has the required properties. This completes the proof of the theorem.

Thus, whereas affine transformations enable us to map any r-simplex into any other (Theorem 7), projective transformations enable us in addition to map any relatively interior point of the first simplex into any relatively interior point of the second.

1.3 THREE BASIC THEOREMS

The three theorems proved in this section are of considerable theoretical importance. The second of these (Helly's Theorem) is included because of its intrinsic interest; it will not be used in the sequel.

Theorem 9. (Radon's Theorem). Let $X = \{x_1,\ldots,x_r\}$ be any finite set of points in E^d. If $r \geq d + 2$, then X can be partitioned into two subsets X_1 and X_2, such that $\text{conv } X_1 \cap \text{conv } X_2 \neq \emptyset$.

The partitioning given by the theorem may not be unique, as we see from the following examples in E^2 (Figure 2).

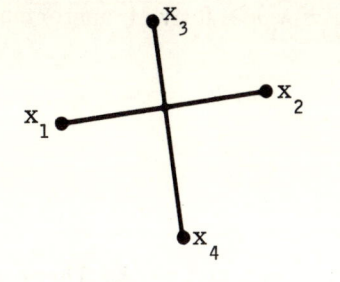

$$\text{conv}\{x_1, x_2\} \cap \text{conv}\{x_3, x_4\} \neq \emptyset.$$

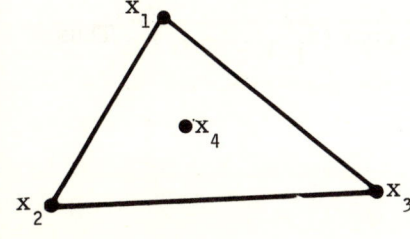

$$\text{conv}\{x_1, x_2, x_3\} \cap \text{conv}\{x_4\} \neq \emptyset.$$

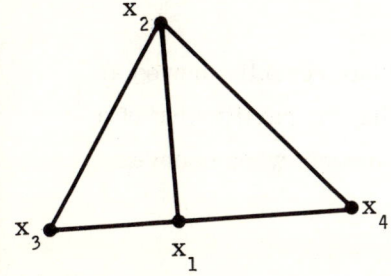

$$\begin{cases} \text{conv}\{x_1, x_2\} \cap \text{conv}\{x_3, x_4\} \neq \emptyset, \\ \text{conv}\{x_1\} \cap \text{conv}\{x_2, x_3, x_4\} \neq \emptyset. \end{cases}$$

Figure 2

Proof. E^d is a d-dimensional affine space, so that any maximal affinely independent set contains $d+1$ points. Since $r \geq d+2$, the points $x_1, \ldots, x_r$ must be affinely dependent, so that we can find scalars $\lambda_1, \ldots, \lambda_r$, not all zero, such that

$$\sum_{i=1}^{r} \lambda_i x_i = 0, \quad \sum_{i=1}^{r} \lambda_i = 0.$$

Without loss of generality, suppose that for some j with $1 \leq j < r$, $\lambda_1 > 0, \ldots, \lambda_j > 0$, $\lambda_{j+1} \leq 0, \ldots, \lambda_r \leq 0$. Put

$$\lambda = \lambda_1 + \ldots + \lambda_j = -(\lambda_{j+1} + \ldots + \lambda_r) > 0,$$

and define x by

$$x = \sum_{i=1}^{j} \frac{\lambda_i}{\lambda} x_i = \sum_{i=j+1}^{r} (-\frac{\lambda_i}{\lambda}) x_i.$$

Then x is a convex combination of $x_1, \ldots, x_j$, and so, by Theorem 4 $x \in \text{conv}\{x_1, \ldots, x_j\}$. Likewise $x \in \text{conv}\{x_{j+1}, \ldots, x_r\}$. Thus, letting $X_1 = \{x_1, \ldots, x_j\}$ and $X_2 = \{x_{j+1}, \ldots, x_r\}$,

$$\text{conv } X_1 \cap \text{conv } X_2 \neq \emptyset,$$

as required. This completes the proof of the theorem.

We mention that Tverberg [10] has recently proved an extension of Radon's Theorem, involving the partitioning of a given set into an arbitrary number of subsets whose convex hulls have non-empty intersection.

Theorem 10. (Helly's Theorem). Let $\{K_1, \ldots, K_r\}$ be a set of r convex sets in E^d, with $r \geq d + 1$, such that every subset of $d + 1$ of these sets has non-empty intersection. Then $\bigcap_{i=1}^{r} K_i \neq \emptyset$.

The assertion is the strongest possible, in the sense that the theorem is no longer true if we allow even one of the given sets to be non-convex. An example of this is illustrated in Figure 3 for the case $d = 2$, $r = 4$.

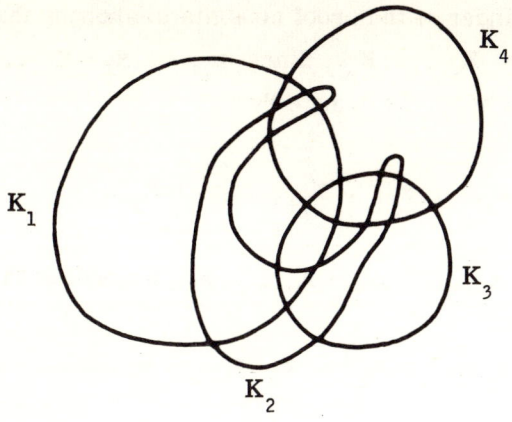

Figure 3

Proof. If $r = d + 1$ then the theorem is trivially true. We prove it for $r > d + 1$ by induction on r, and make the induction hypothesis that the theorem is true for all families of $r - 1$ convex sets. This implies that, for $i = 1, \ldots, r$, there exists a point x_i such that

$$x_i \in K_1 \cap \ldots \cap K_{i-1} \cap K_{i+1} \cap \ldots \cap K_r .$$

Since $r \geq d + 2$, we can apply Radon's Theorem 9 to the set $X = \{x_1, \ldots, x_r\}$, and hence find two subsets X_1 and X_2 of X such that $\text{conv } X_1 \cap \text{conv } X_2 \neq \emptyset$. Without loss of generality, let $X_1 = \{x_1, \ldots, x_j\}$, $X_2 = \{x_{j+1}, \ldots, x_r\}$, and then we can find a point x such that

$$x \in \text{conv}\{x_1, \ldots, x_j\} \cap \text{conv}\{x_{j+1}, \ldots, x_r\} .$$

25

The remainder of the proof consists in showing that x lies in each of the sets $K_1, \ldots, K_r$. Since $x_1, \ldots, x_j \in K_{j+1}, \ldots, K_r$, and each K_i is convex, we deduce that

$$x \in \text{conv}\{x_1, \ldots, x_j\} \subseteq K_{j+1} \cap \ldots \cap K_r .$$

Similarly, since $x_{j+1}, \ldots, x_r \in K_1, \ldots, K_j$, we deduce that

$$x \in \text{conv}\{x_{j+1}, \ldots, x_r\} \subseteq K_1 \cap \ldots \cap K_j .$$

Thus $x \in \bigcap_{i=1}^{r} K_i$, and so the theorem is proved.

Theorem 10 is the 'finite form' of Helly's Theorem. The statement of the theorem remains true for infinite families of convex sets, as long as they are all closed, and at least one is bounded.

The final theorem of this section strengthens Theorem 4. That theorem asserted that the convex hull of a set can be constructed by taking convex combinations of finite subsets; here we assert that in E^d subsets of at most $d + 1$ points are sufficient.

Theorem 11. (Carathéodory's Theorem). <u>The convex hull of a subset</u> X <u>of</u> E^d <u>is precisely the set of all convex combinations of subsets of</u> X <u>containing at most</u> $d + 1$ <u>points.</u>

Proof. Let x be any point of conv X. By Theorem 4, we can write

$$x = \sum_{i=1}^{r} \lambda_i x_i \quad (x_i \in X, \quad \lambda_i \geq 0, \quad \sum_{i=1}^{r} \lambda_i = 1) .$$

Suppose that this is a minimal representation of x, in the sense that x cannot be expressed as a convex combination of fewer than r points of X. We shall show that the hypothesis $r \geq d + 2$ leads to a contradiction. In E^d, at most $d + 1$ points can be affinely independent, and so there exists a non-trivial affine relation

$$\sum_{i=1}^{r} \mu_j x_i = 0, \quad \sum_{i=1}^{r} \mu_i = 0.$$

We now subtract an appropriate multiple of this relation from the expression above for x, to eliminate one of the x_i. Precisely, consider all the positive μ_i (there must be at least one), and select that suffix i (say i_0) for which $\dfrac{\lambda_i}{\mu_i}$ takes its minimal (positive) value. Then

$$x = \sum_{i=1}^{r} \lambda_i x_i = \sum_{i=1}^{r} (\lambda_i - \dfrac{\lambda_{i_0}}{\mu_{i_0}} \mu_i) x_i .$$

This is a convex combination of (strictly) less than r points of X, which contradicts our original assumption. Hence in fact $r \leq d + 1$, and the theorem is proved.

Carathéodory's Theorem has the following consequences.

Corollary 1. <u>For any set $X \subseteq E^d$, conv X is the union of all the simplices which are convex hulls of subsets of X.</u>

Corollary 2. <u>Every polytope is a finite union of simplices.</u>

Another consequence, which we require later, is the following:

Corollary 3. _If_ X _is a closed bounded set in_ E^d, _then_ conv X _is also closed and bounded._

Proof. Define a mapping $\Delta : E^{(d+1)^2} \to E^d$ as follows. A point of $E^{(d+1)^2}$ can be written in the form

$$(\lambda_0, \ldots, \lambda_d, x_0, \ldots, x_d)$$

where $\lambda_0, \ldots, \lambda_d$ are real numbers, and $x_0, \ldots, x_d \in E^d$. Let

$$(\lambda_0, \ldots, \lambda_d, x_0, \ldots, x_d) \Delta = \sum_{i=0}^{d} \lambda_i x_i \;;$$

clearly Δ is continuous.

Let

$$X_1 = \{(\lambda_0, \ldots, \lambda_d, x_0, \ldots, x_d) \in E^{(d+1)^2} \mid \lambda_i \geq 0, \sum_{i=0}^{d} \lambda_i = 1, \; x_i \in X\}.$$

Since X is closed and bounded, X_1 is also closed and bounded. By Carathéodory's Theorem, $X_1 \Delta = \text{conv } X$, and as Δ is continuous, conv X is closed and bounded, which proves the corollary.

1.4   SUPPORT PROPERTIES

We shall now discuss the support properties of a closed bounded convex set. This is in preparation for a discussion of the facial structure of a convex set, and particularly a convex polytope, in the next chapter.

The set of points lying on, or to one side of, a hyperplane H in E^d is called a closed half-space; the set of points strictly to one side of a hyperplane is called an open half-space. If

$$H = \{x \in E^d \mid \langle x, a \rangle = \beta\} \quad (a \neq 0),$$

then the two closed half-spaces determined by H are

$$H^+ = \{x \in E^d \mid \langle x, a \rangle \geq \beta\},$$

$$H^- = \{x \in E^d \mid \langle x, a \rangle \leq \beta\}.$$

If we replace the inequalities in the definitions of H^+ and H^- by strict inequalities, then we obtain the corresponding open half-spaces.

Let K be a closed bounded convex set in E^d. Then a hyperplane H of E^d is said to support K if $H \cap K \neq \emptyset$, and K is contained in one of the closed half-spaces determined by H. If $x \in H \cap K$, then we say H supports K at x. H is also called a supporting hyperplane of K. That closed half-space determined by H which contains K is called a supporting half-space of K. If $K \subseteq H^+$, then -a is called an outward normal vector of H; if $K \subseteq H^-$, then a is an outward normal vector (see Figure 4).

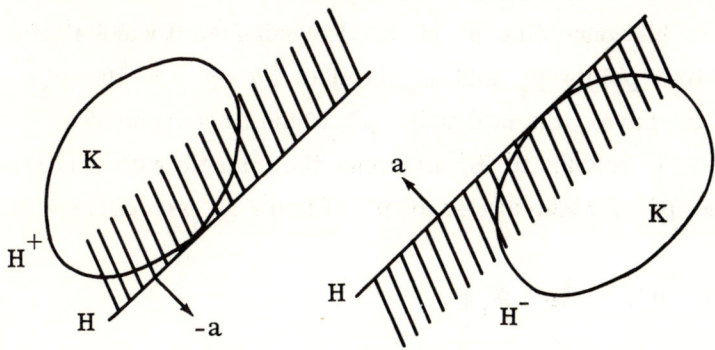

Figure 4

The first result is almost trivial.

Theorem 12. <u>Let K be a closed bounded convex set in E^d, and let a be a non-zero vector. Then there exists a supporting hyperplane of K with outward normal a.</u>

Proof. The required hyperplane is

$$H = \{x \in E^d \mid \langle x, a \rangle = \sup_{y \in K} \langle y, a \rangle \}.$$

The supremum is attained, since K is closed and bounded.

Our next objective is to show that if K is any closed bounded convex set in E^d, then through every point $x \in \operatorname{bd} K$ (the boundary of K) there is a supporting hyperplane of K. In order to prove this result (Theorem 14), we introduce the useful concept of the nearest point map for K, which is defined as follows.

Let $p \in E^d \setminus K$ be a given point. Since K is closed and bounded, the infimum

$$\inf_{x \in K} \|x - p\|$$

is attained, and is finite and strictly positive. The infimum is attained for just one point p' of K. For suppose it were attained at two distinct points p'_1 and p'_2, so that $\|p - p'_1\| = \|p - p'_2\|$. Since K is convex, the mid-point p'' of the line-segment $\operatorname{conv}\{p'_1, p'_2\}$ belongs to K, and since the triangle with vertices p, p'_1 and p'' is right-angled at p'' (Figure 5), we conclude that

$$\|p - p''\| < \|p - p'_1\|,$$

which contradicts the definition of p'_1 (and p'_2).

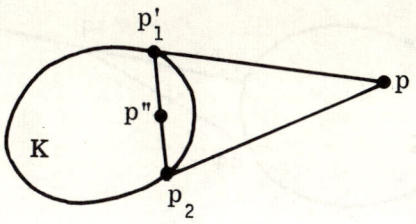

Figure 5

We deduce that the map $\Phi : E^d \to K$, given by

$$p\Phi = \begin{cases} p & \text{for } p \in K, \\ p' & \text{for } p \in E^d \setminus K, \end{cases}$$

is well-defined. Φ is called the <u>nearest-point map</u> for K.

Let $p \in E^d \setminus K$, so that $p\Phi \neq p$. Then the open half-line containing p with end-point $p\Phi$ is called a <u>ray</u>, and is denoted by $R(p\Phi, p)$. Rays have the following properties.

Lemma 1. <u>If</u> $q \in R(p\Phi, p)$, <u>then</u> $q\Phi = p\Phi$.

Proof. If $q = p$, the result is trivial. If $q \in \text{conv}\{p, p\Phi\}$, and $q\Phi \neq p\Phi$, then (Figure 6)

$$\|p - q\Phi\| \leq \|p - q\| + \|q - q\Phi\|$$
$$< \|p - q\| + \|q - p\Phi\|$$
$$= \|p - p\Phi\|,$$

which contradicts the definition of $p\Phi$.

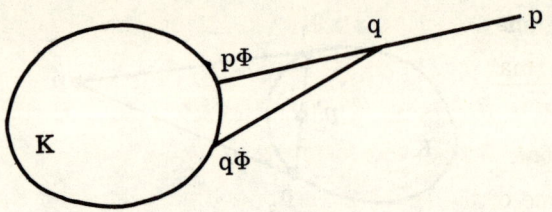

Figure 6

If $q \notin \text{conv}\{p, p\Phi\}$, then $p \in \text{conv}\{q, p\Phi\}$. Let $q\Phi \neq p\Phi$, and let p'' be the point in which the line through p parallel to $q\Phi - p$ meets the line segment $\text{conv}\{p\Phi, q\Phi\}$ (Figure 7). Then since $\|q - q\Phi\| < \|q - p\Phi\|$, from consideration of the similar triangles with vertices $\{p\Phi, p, p''\}$ and $\{p\Phi, q, q\Phi\}$, we deduce that

$$\|p - p''\| < \|p - p\Phi\|,$$

which contradicts the definition of $p\Phi$. Hence in this case also, $q\Phi = p\Phi$, and the lemma is proved.

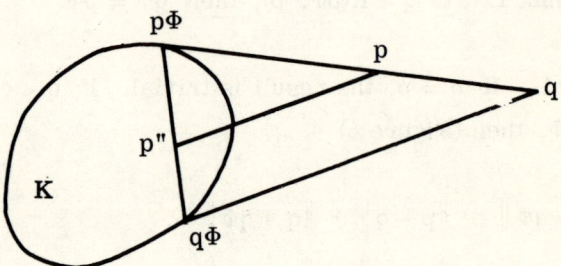

Figure 7

Lemma 2. Let $p \in E^d \setminus K$. Then the hyperplane H through $p\Phi$ and normal to $R(p\Phi, p)$ supports K.

Proof. Since $p\Phi \in H \cap K$, it is only necessary to show that K lies in one of the two closed half-spaces bounded by H: in fact we shall show that K lies in the half-space that does not contain p.

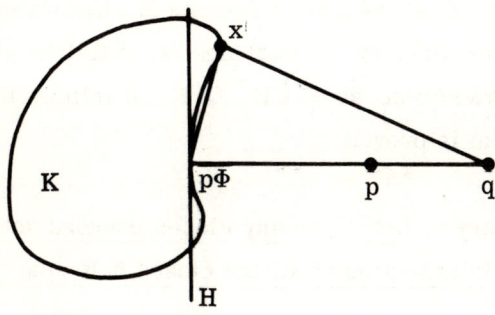

Figure 8

For suppose this is not so, and we can find a point $x \in K$ lying on the same side of H as p. Then the hyperplane through x, normal to $p\Phi - x$, will meet $R(p\Phi, p)$ in a point q (Figure 8). Now by Lemma 1, $p\Phi = q\Phi$, and since the triangle with vertices x, $p\Phi$ and q is right-angled at x,

$$\|q - x\| \leq \|q - p\Phi\| = \|q - q\Phi\|,$$

contradicting the definition of $q\Phi$. Hence no such point x can exist, which proves the lemma.

Theorem 13. <u>Every closed bounded convex set K in E^d is the intersection of all its supporting half-spaces.</u>

Proof. Let H^+ represent a general supporting half-space of K, and write $\bigcap H^+$ for the intersection of all the supporting half-spaces. Since each $H^+ \supseteq K$, we deduce that $\bigcap H^+ \supseteq K$. Let us suppose that there exists $x \in \bigcap H^+ \setminus K$. Then x and its image $x\Phi$ under the nearest point map determine a hyperplane H which supports K at $x\Phi$, as in Lemma 2. Because the supporting half-space determined by H excludes x, it follows that $x \notin \bigcap H^+$. This is a contradiction, so $\bigcap H^+ \subseteq K$. It follows that $\bigcap H^+ = K$, and the theorem is proved.

Corollary. <u>Let S be any closed bounded set in E^d. Then conv S is the intersection of all the closed half-spaces containing S.</u>

This follows immediately from the fact that because each half-space H^+ is convex, if $H^+ \supseteq S$, then $H^+ \supseteq \text{conv } S$. But conv S is closed and bounded by Corollary 3 of Theorem 11. Hence Theorem 13 implies that $\text{conv } S = \bigcap H^+$, and so the result is proved.

It is worth noticing that this result is not generally true if S is not bounded. For example, in E^2, let

$$S = \{(\pm n, n^{-1}) \mid n = 1, 2, 3, \ldots \}.$$

Then conv S is not closed, so it cannot be the intersection of a family of <u>closed</u> half-spaces.

Lemma 3. (Busemann-Feller Lemma). For any closed bounded convex set K, the nearest point map Φ does not increase length; that is, for all p, q ∈ E^d,

$$\|p - q\| \geq \|p\Phi - q\Phi\|.$$

Proof. We give a proof in the case p, q ∈ $E^d \setminus K$; the case where just one of p and q belongs to K is similar, and the remaining case is trivial. (Figure 9 illustrates the case d = 2, but the method of proof holds for all dimensions.)

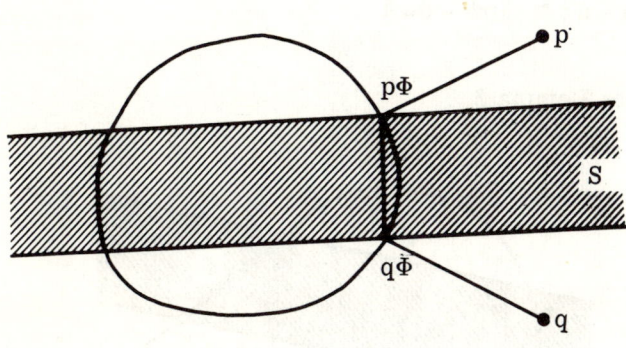

Figure 9

If pΦ = qΦ, then the result is again trivial, so we assume pΦ ≠ qΦ. Let S be the open strip bounded by parallel hyperplanes $H_{p\Phi}$ and $H_{q\Phi}$ passing through pΦ and qΦ respectively, and normal to the vector pΦ - qΦ. That is, if ⟨ qΦ, pΦ - qΦ ⟩ < ⟨ pΦ, pΦ - qΦ ⟩,

$$S = \{x \in E^d | \langle q\Phi, p\Phi - q\Phi \rangle < \langle x, p\Phi - q\Phi \rangle < \langle p\Phi, p\Phi - q\Phi \rangle \}.$$

If p lay in the open half-space bounded by $H_{p\Phi}$ and containing S (Figure 10) then $R(p\Phi, p)$ would meet $H_{q\Phi}$ in some point x, and since the triangle with vertices $p\Phi$, $q\Phi$ and x is right-angled at $q\Phi$, clearly

$$\|x - q\Phi\| < \|x - p\Phi\|.$$

Since $x\Phi = p\Phi$ by Lemma 1, this is a contradiction. We deduce that $p \notin S$, and similarly $q \notin S$. But $H_{p\Phi}$ and $H_{q\Phi}$ are at a distance $\|p\Phi - q\Phi\|$ apart, and so

$$\|p - q\| \geq \|p\Phi - q\Phi\|.$$

This proves Lemma 3.

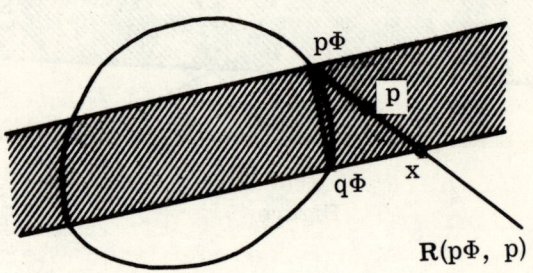

Figure 10

Lemma 4. <u>The nearest point map $\Phi : E^d \to K$ is continuous.</u>

Proof. This is an immediate consequence of Lemma 3, which shows that Φ satisfies a Lipschitz condition.

Lemma 5. Let S^{d-1} be a $(d-1)$-sphere bounding an open d-ball which contains the closed bounded convex set K in E^d. Then the nearest point map Φ (for K) maps S^{d-1} onto the boundary bd K of K.

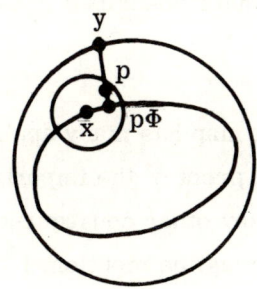

Figure 11

Proof. Let $x \in$ bd K. For each $\eta > 0$, the open d-ball with centre x and radius η contains a point $p \in E^d \setminus K$, and since $x\Phi = x$, by Lemma 3

$$\|x - p\Phi\| \leq \|x - p\| < \eta \ .$$

Let $R(p\Phi, p)$ meet S^{d-1} in y (which is uniquely determined), so that $y\Phi = p\Phi$. Now let $\eta \to 0$. We obtain a sequence of points y_i on S^{d-1}, and the corresponding sequence of points $y_i \Phi \to x$. Since S^{d-1} is compact, there exists a convergent subsequence (y'_j), which converges to some point $q \in S^{d-1}$. By Lemma 4, Φ is continuous, so that $q\Phi = x$. Hence each point $x \in$ bd K is the image of some point $q \in S^{d-1}$, and so the lemma is proved.

From these lemmas, the main theorem of the section follows immediately.

Theorem 14. <u>Through every point</u> x <u>on the boundary of a closed bounded convex set</u> K <u>passes a supporting hyperplane</u> H.

Proof. By Lemma 5, every point $x \in \text{bd } K$ can be written $x = q\Phi$ for some $q \in E^d \setminus K$. Then the hyperplane constructed from q and $q\Phi = x$ in the manner described in Lemma 2 contains x and supports K.

The nearest point map has many applications in the theory of convexity, such as the proof of the important Theorem 14 above. We cannot discuss its many other consequences here, but the following basic theorem must be mentioned.

Theorem 15. <u>Every closed bounded convex set</u> K <u>has a volume in the Peano-Jordan sense.</u>

Proof. Let B^d be any closed d-ball in E^d containing K. Then since $K \subseteq B^d$, it follows that K is the image of B^d under the continuous map Φ (Lemma 4). Since B^d has a volume, so has K.

2. Convex Polytopes

2.1 THE FACES OF A CONVEX POLYTOPE

The combinatorial theory of convex polytopes is largely concerned with their 'facial structure'. In this section we shall establish the basic theorems concerning the faces of a polytope.

Definition. Let H be a supporting hyperplane of a closed bounded convex set K. Then H ∩ K is called a <u>face</u> of K.

For technical reasons it is sometimes convenient also to include among the faces of K, two further faces, namely the empty set ∅ and K itself. These are called <u>improper</u> faces of K; all other faces are said to be <u>proper</u>. In the following it will be clear from the context whether improper faces are being considered as faces of K.

Every face F of K is convex, being the intersection of two convex sets. If dim F = j, then F is called a j-<u>face</u> of K. Thus if K is d-dimensional, and F is a proper face of K, then $0 \leq \dim F \leq d - 1$. Recalling the convention for the dimension of the empty set given in §1.1, it is natural to regard the improper faces as corresponding to the dimensions j = -1 (for ∅) and j = d (for K). The 0-faces of K are called <u>vertices,</u> and the set of vertices of K (regarded as points) is denoted by vert K. The 1-faces of K are called <u>edges</u>, and the (d - 1)-faces are called <u>facets</u> of K. Thus facets of K are proper faces of the greatest possible dimension.

The faces of polytopes have a number of properties that do not hold in the case of general (closed bounded) convex sets. Some of these are established in Theorems 1, 2 and 5.

Theorem 1. <u>A polytope has only a finite number of distinct faces, and each face is a convex polytope.</u>

Proof. Suppose that the polytope P is the convex hull of the finite set $\{x_1, \ldots, x_r\}$, and that the equation of the supporting hyperplane H is $\langle x, a \rangle = \gamma$. Without loss of generality we may suppose that $\{x_1, \ldots, x_j\} \subset H$, and that $x_{j+1}, \ldots, x_r$ lie in the open half-space

$$\operatorname{int} H^+ = \{x \in E^d | \langle x, a \rangle > \gamma\}.$$

Then $\langle x_i, a \rangle = \gamma$ $(i = 1, \ldots, j)$, and $\langle x_i, a \rangle = \gamma + \delta_i$ $(i = j+1, \ldots, r)$, for some $\delta_i > 0$. Let

$$x = \sum_{i=1}^{r} \lambda_i x_i, \quad \lambda_i \geq 0, \quad \sum_{i=1}^{r} \lambda_i = 1,$$

be an arbitrary point of P. Then

$$\langle x, a \rangle = \sum_{i=1}^{r} \lambda_i \langle x_i, a \rangle = (\sum_{i=1}^{r} \lambda_i)\gamma + \sum_{i=j+1}^{r} \lambda_i \delta_i$$

$$= \gamma + \sum_{i=j+1}^{r} \lambda_i \delta_i.$$

The point x lies in H if and only if $\langle x, a \rangle = \gamma$; that is, if and only if $\sum_{i=j+1}^{r} \lambda_i \delta_i = 0$. Since each δ_i is strictly positive, and since each λ_i is non-negative, a necessary and sufficient condition for this is $\lambda_{j+1} = \ldots = \lambda_r = 0$; in other words, that x is a convex combination of $x_1, \ldots, x_j$ only. Hence

$$H \cap P = \text{conv } \{x_1, \ldots, x_j\},$$

and so $H \cap P$ is a polytope. This proves the second statement of the theorem.

Since the set $\{x_1, \ldots, x_r\}$ is finite, it has only a finite number of subsets, and since each face of P corresponds to a subset of $\{x_1, \ldots, x_r\}$, it follows that P has only a finite number of distinct faces. The first statement of the theorem is therefore also true.

Theorem 2. <u>A convex polytope P is the convex hull of its set of vertices; that is,</u>

$$P = \text{conv (vert } P).$$

Proof. Suppose $P = \text{conv}\{x_1, \ldots, x_r\}$. From the definition of the convex hull, it is immediately clear that if

$$x_i \in \text{conv}\{x_1, \ldots, x_{i-1}, x_{i+1}, \ldots, x_r\}$$

for any $i = 1, \ldots, r$, then

$$P = \text{conv}\{x_1, \ldots, x_{i-1}, x_{i+1}, \ldots, x_r\}.$$

We may therefore successively remove each point x_i that lies in the convex hull of the remaining points, eventually obtaining a representation

$$P = \text{conv}\ \{x_1, \ldots, x_v\}$$

say, of P as the convex hull of a minimal set of points. The remainder of the proof consists in showing that

$$\{x_1, \ldots, x_v\} = \text{vert}\ P\ .$$

Consider any point x_i ($1 \le i \le v$), and let

$$P_i = \text{conv}\ \{x_1, \ldots, x_{i-1}, x_{i+1}, \ldots, x_v\}.$$

Then $x_i \in E^d \backslash P_i$, and so there is a nearest point $x_i \Phi \in P_i$ to x_i. By Lemma 2 of §1.4, the hyperplane H' through $x_i \Phi$ with normal $x_i \Phi - x_i$ supports K, and therefore the points $x_1, \ldots, x_{i-1}, x_{i+1}, \ldots, x_v$ all lie in the closed half-space bounded by H' which does not contain x_i (see Figure 12).

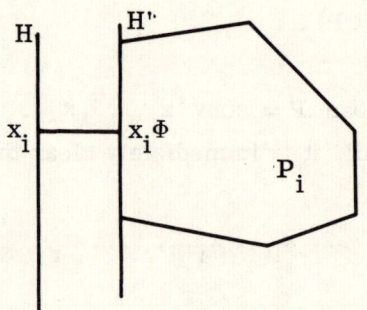

Figure 12

Let H be the hyperplane through x_i parallel to H'. Then $x_1, \ldots, x_v$ all lie in one of the closed half-spaces H^+ bounded by H, and so $P \subset H^+$. Further, it is clear that only one of the points $x_1, \ldots, x_v$ lie in H, namely x_i, and so by Theorem 1,

$$H \cap P = \mathrm{conv}\{x_i\} = \{x_i\}.$$

Hence H supports P, and the point x_i is a face (vertex) of P. This completes the proof of Theorem 2.

In Chapter 1, Theorem 13, we showed that a closed bounded convex set is the intersection of a family of closed half-spaces. Theorem 3 will show that, in the case of polytopes, a finite family suffices. Theorem 4 is the converse of Theorem 3. These two theorems confirm, in fact, one's intuitive picture of a convex polytope. It is convenient to introduce the following terminology.

Definition. A <u>polyhedral set</u> is the intersection of a finite number of closed half-spaces.

A polyhedral set may, of course, be unbounded.

Theorem 3. <u>A convex polytope is a bounded polyhedral set.</u>

Proof. Suppose, without loss of generality, that P is a d-polytope in E^d. Let $\mathrm{vert}\, P = \{x_1, \ldots, x_v\}$, and let $F_1, \ldots, F_s$ denote the facets of P. Then associated with each facet F_i is a supporting hyperplane H_i (such that $H_i \cap P = F_i$) and a corresponding supporting half-space H_i^+ (bounded by H_i), such that $P \subset H_i^+$. The proof of the theorem depends upon proving that

$$P = H_1^+ \cap \ldots \cap H_s^+ . \tag{1}$$

Firstly, since $H_i^+ \supset P_i$ for each i, we see that

$$P \subseteq H_1^+ \cap \ldots \cap H_s^+ . \tag{2}$$

To prove the opposite inclusion, let us suppose that there exists a point $x \in H_1^+ \cap \ldots \cap H_s^+$, with $x \notin P$. Let D be the union of the affine subspaces of E^d spanned by at most d points, one of which is x, and the remaining d - 1 belong to vert P. Then D consists of a finite union of affine subspaces of dimension at most d - 1, and because dim P = d, this implies that int $P \nsubseteq D$. Hence we can find a point $y \in \text{int } P \setminus D$.

Since $y \in \text{int } P$, $x \notin P$, as we saw in Chapter 1 there exists a unique point $z \in \text{bd } P$ such that

$$\text{conv } \{x, y\} \cap \text{bd } P = \{z\} .$$

We shall show that z belongs to a facet of P, and to no face of lower dimension. For suppose z belonged to a j-face of P $(0 \leq j \leq d - 2)$. Then by Carathéodory's Theorem (Chapter 1, Theorem 11), z lies in some simplex of dimension $r \leq j$, whose vertices $\{w_0, \ldots, w_r\}$ are vertices of P. Thus

$$z \in \text{conv}\{w_0, \ldots, w_r\} ,$$

and so, because r < d - 1, we deduce that $z \in D$. As $x \in D$, this implies that $y \in D$ also, which is a contradiction.

So, let the facet to which z belongs be F_i. Then $z \in H_i$, and since $y \in \text{int } P \subset H_i^+$, we deduce that $x \notin H_i^+$. This contra-

44

dicts our initial assumption that $x \in H_1^+ \cap \ldots \cap H_s^+$, and so we deduce that

$$P \supseteq H_1^+ \cap \ldots \cap H_s^+ .$$

This inclusion, along with (2), implies (1), and since P is bounded, the theorem is proved.

Theorem 4. <u>A bounded polyhedral set is a polytope.</u>

Proof. As usual, we denote by H^+ one of the closed half-spaces bounded by a hyperplane H. Let $Q = \bigcap_{i=1}^{n} H_i^+$ be a given bounded polyhedral set in E^d; without loss of generality we may suppose that Q is d-dimensional. Let us also suppose that none of the half-spaces H_i^+ is redundant, in the sense that $Q \neq \bigcap_{i \neq j} H_i^+$ for any $j = 1, \ldots, n$.

For each $j = 1, \ldots, n$, let $F_j = H_j \cap Q$. Then F_j is a convex set in H_j, whose interior, relative to H_j, is $H_j \cap \text{int} \bigcap_{i \neq j} H_i^+$. This set is non-empty, for if $H_j \cap \text{int} \bigcap_{i \neq j} H_i^+ = \emptyset$, then since a convex set is connected, $\text{int} \bigcap_{i \neq j} H_i^+ \subset H_j^+$ or H_j^-. The latter is impossible, and the former implies that $\bigcap_{i \neq j} H_i^+ \subseteq H_j^+$, which means that, contrary to hypothesis, H_j^+ is redundant. We deduce that F_j is (d - 1)-dimensional, and so is a facet of Q.

Every point of bd Q lies on the boundary of one of the half-spaces H_j^+ (that is, it belongs to H_j for some j); it thus follows that

$$\text{bd } Q = \bigcap_{j=1}^{n} F_j .$$

45

Also, since

$$F_j = H_j \cap Q = \bigcap_{i \neq j} (H_j \cap H_i^+)$$

and since, for $i = 1, \ldots, n$, $H_j \cap H_i^+$ is either the whole of H_j, or a closed half-space in H_j, we see that F_j is itself a bounded polyhedral set in H_j.

We now use induction on the dimension. The theorem clearly holds if $d = 1$, so assume that it holds in $d - 1$ or fewer dimensions. Since $\dim F_j = d - 1$, it follows that F_j is a polytope, and so by Theorem 2,

$$F_j = \text{conv (vert } F_j) .$$

Now let V be the finite set of points.

$$V = \bigcup_{j=1}^{n} \text{vert } F_j .$$

Since $V \subseteq Q$, and Q is convex, we conclude that

$$\text{conv } V \subseteq Q .$$

We can also show that $Q \subseteq \text{conv } V$. For any point $x \in Q$, there are two possibilities:

(1) $x \in \text{bd } Q$. Then as we have seen $x \in F_j$ for some j, and since vert $F_j \subseteq V$, this implies that $x \in \text{conv } V$.

(2) $x \in \text{int } Q$. Let L be any line through x. Then L meets bd Q in two points (since Q is convex), say x_0 and x_1. Now $x_0 \in \text{bd } Q$, and so, as in (1), $x_0 \in \text{conv } V$. Similarly

$x_1 \in \text{conv } V$, and since $x \in \text{conv}\{x_0, x_1\}$, we deduce that $x \in \text{conv } V$.

Hence in either case $x \in \text{conv } V$, and since x was a general point of Q, we have $Q \subseteq \text{conv } V$. Thus $Q = \text{conv } V$ is the convex hull of a finite set of points, and so Q is a convex polytope. This completes the proof of Theorem 4.

The following corollaries are direct consequences of Theorems 3 and 4.

Corollary 1. <u>If a d-polytope P in E^d has n facets, then it can be written as the intersection of n closed half-spaces.</u>

Corollary 2. <u>If P is a convex polytope in E^d, and L is any affine subspace of E^d, then $L \cap P$ is also a convex polytope.</u>

For, if H^+ is any closed half-space in E^d, then $L \cap H^+$ is either L itself, or a closed half-space in L. Since P is a bounded polyhedral set (Theorem 3), so is $L \cap P$, and therefore $L \cap P$ is a polytope (Theorem 4).

Theorem 5. <u>Let K_1, K_2 be closed bounded convex sets, such that $K_2 \subseteq K_1$. If F is a face of K_1, then $F \cap K_2$ is a face of K_2.</u>

Proof. If F is an improper face of K_1, there is nothing to prove. Otherwise, let H be a supporting hyperplane of K_1, such that $H \cap K_1 = F$. Clearly, either $H \cap K_2 = \emptyset$ or H supports K_2. In the former case, $H \cap K_2 = F \cap K_2 = \emptyset$ is a face of K_2. In the latter, $H \cap K_2 = F \cap K_2$ is again a face of K_2, which completes the proof of the theorem.

Corollary. *Let F_1, F_2 be faces of a closed bounded convex set, such that $F_2 \subseteq F_1$. Then F_2 is a face of F_1.*

This is an immediate consequence of the theorem.

If K is a closed bounded d-dimensional convex set in E^d, then Theorem 14 of Chapter 1, together with the definition of a face, implies that bd K is the union of all the proper faces of K. From the proof of Theorem 4, we see that in the case of a d-polytope P, bd P is the union of all the facets of P, and that there is no need to include any face of dimension less than d - 1. The next result follows at once from this observation.

Theorem 6. *Let F be any proper face of a polytope P. Then F is a face of some facet of P.*

Proof. Choose any point $x \in$ relint F. Then $x \in$ bd P (since $F \subset$ bd P), and so there is some facet F_1 of P such that $x \in F_1$. Let H_1 be any supporting hyperplane of P such that $H_1 \cap P = F_1$. Since $x \in$ relint $F \cap H_1$, and H_1 supports P, it is clear that $F \subset H_1$. Hence $F \subseteq F_1$, and by Theorem 5 this implies that F is a face of F_1, which proves the theorem.

Corollary. *Let F^j, F^k be j- and k- faces respectively of a polytope P, such that $F^j \subseteq F^k$. Then there are faces F^i (i = j + 1, ..., k - 1) of P, such that*

$$F^j \subset F^{j+1} \subset \ldots \subset F^{k-1} \subset F^k.$$

For by Theorem 5, F^j is a face of F^k, and if F^j is a proper face of F^k, then by Theorem 6, F^j is a face of some facet,

F^{k-1} say, of F^k. The corollary follows at once by induction on k.

Theorem 5 shows that if F_1 and F_2 are faces of a polytope P, such that $F_2 \subseteq F_1$, then F_2 is a face of F_1. Theorem 7 is the converse result.

Theorem 7. <u>Let F_1 be a face of a polytope P, and let F_2 be a face of the polytope F_1. Then F_2 is a face of P.</u>

Thus the relation 'being a face of' is transitive on the set of all faces of a polytope: it is not so for general (closed bounded) convex sets. For example, consider the set K which consists of two half-discs adjoined to a rectangle $\text{conv}\{x_1, x_2, x_3, x_4\}$ (see Figure 13).

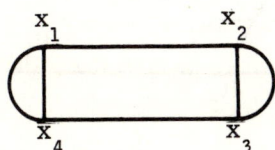

Figure 13

Then $F = \text{conv}\{x_1, x_2\}$ is a 1-face of K, and $\{x_1\}$ is a 0-face of F, but $\{x_1\}$ is not a face of K.

Proof. There are several possible proofs of Theorem 7, including one based on the alternative definition of a polytope given by Theorems 3 and 4. We have chosen a direct proof, whose basic idea is very simple. Since F_1 is a face of P, there is a supporting hyperplane H_1 of P, such that $H_1 \cap P = F_1$. Since F_2 is a face of F_1, there exists a supporting hyperplane H_2 of F_1 in H_1, such that $H_2 \cap F_1 = H_2 \cap P = F_2$. (See Figure 14.) Then it is intuitively clear that if we 'tilt' H_1 about H_2 through a sufficiently small angle, the tilted hyperplane will intersect P in F_2 only. A rigorous proof can be constructed along these lines; however, the algebra is easier if, instead, we tilt H_1 about H_2 through P beyond the point at which it next supports P.

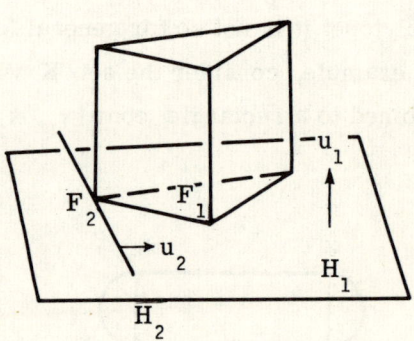

Figure 14

Without loss of generality let us suppose that P is a d-polytope in E^d, that all faces concerned are proper faces, and that the origin $o \in F_2$. Since $o \in F_2 \subset F_1$, the supporting hyperplane H_1 of P such that $H_1 \cap P = F_1$ can be written

$$H_1 = \{x \in E^d | \langle x, u_1 \rangle = 0\},$$

where u_1 is a unit vector, and

$$P \subset H_1^+ = \{x \in E^d | \langle x, u_1 \rangle \geq 0\}.$$

(Thus u_1 is the inward normal vector to H_1.) Since $o \in F_2$, and F_2 is a face of F_1, there is a supporting hyperplane H_2 of F_1 in H_1, such that $H_2 \cap F_1 = F_2$, namely

$$H_2 = \{x \in H_1 | \langle x, u_2 \rangle = 0\},$$

where u_2 is a unit vector in H_1, and

$$F_1 \subset H_2^+ = \{x \in H_1 | \langle x, u_2 \rangle \geq 0\}.$$

Now write

$$H(\eta) = \{x \in E^d | \langle x, \eta u_1 + u_2 \rangle = 0\}.$$

Then clearly $H(\eta) \supset H_2 \supset F_2$ for every η. We shall show that η can be chosen in such a way that $H(\eta) \cap P = F_2$, and $P \subset H(\eta)^+$, where

$$H(\eta)^+ = \{x \in E^d | \langle x, \eta u_1 + u_2 \rangle \geq 0\}$$

is one of the closed half-spaces bounded by $H(\eta)$.

Write

$$\eta_0 = \min\left\{-\frac{\langle w, u_2 \rangle}{\langle w, u_1 \rangle} \middle| w \in \text{vert } P \setminus \text{vert } F_1 \right\},$$

noticing that, since $P \subset H_1^+$, $\langle w, u_1 \rangle > 0$ for all $w \in \text{vert } P \setminus \text{vert } F_1$. Then if $\eta > \eta_0$, it follows that

(i) if $w \in \text{vert } P \setminus \text{vert } F_1'$, then

$$\langle w, \eta u_1 + u_2 \rangle > \eta_0 \langle w, u_1 \rangle + \langle w, u_2 \rangle \geq 0.$$

(ii) if $w \in \text{vert } F_1 \setminus \text{vert } F_2$, then

$$\langle w, \eta u_1 + u_2 \rangle = \langle w, u_2 \rangle > 0.$$

(iii) if $w \in \text{vert } F_2$, then

$$\langle w, \eta u_1 + u_2 \rangle = 0.$$

Thus all the vertices of F_2 lie in $H(\eta)$, and the remaining vertices of P lie in the interior of $H(\eta)^+$. Hence $H(\eta)$ supports P, with $H(\eta) \cap P = F_2$, so that F_2 is a face of P, as was to be proved.

It may be noticed that $H(\eta_0)$ is also a supporting hyperplane of P; however, $H(\eta_0)$ will contain vertices of P apart from those of F_2.

Theorem 8. <u>Let $\{F_1, \ldots, F_r\}$ be a family of faces of a closed bounded convex set K. Then $\bigcap_{i=1}^{r} F_i$ is also a face of K.</u>

Proof. Write $F = \bigcap_{i=1}^{r} F_i$. If $F = \emptyset$, or $r = 1$ then the result is trivially true, so, let us assume that $F \neq \emptyset$ and $r > 1$. Without loss of generality we may take each face F_i to be a proper face of K, and choose our coordinate system in E^d so that $o \in F$. For $i = 1, \ldots, r$ there is a supporting hyperplane H_i of K such that $H_i \cap K = F_i$, and since $o \in F_i$, we can write

$$H_i = \{x \in E^d | \langle x, u_i \rangle = 0\}$$

where u_i is a unit vector, and

$$K \subset H_i^+ = \{x \in E^d | \langle x, u_i \rangle \geq 0\}.$$

Let H be the hyperplane

$$H = \{x \in E^d | \langle x, w \rangle = 0\},$$

where $w = \sum_{i=1}^{r} u_i$. Clearly

$$K \subset H^+ = \{x \in E^d | \langle x, w \rangle \geq 0\},$$

and since $o \in H \cap K$, this implies that H supports K. Now if $x \in F$, then $\langle x, u_i \rangle = 0$ ($i = 1, \ldots, r$), and so $x \in H \cap K$. Thus $F \subseteq H \cap K$. On the other hand, if $x \in K \backslash F$, then $\langle x, u_i \rangle > 0$ for at least one i, and so $\langle x, w \rangle > 0$. Hence $x \notin H \cap K$, which shows that $H \cap K \subseteq F$. That is, $F = H \cap K$, and so F is a face of K, as claimed.

We remark that Theorem 8 is, in fact, true for arbitrary families of faces of K.

If F_j is a facet of a d-polytope P in E^d, and H_j is the supporting hyperplane of P such that $H_j \cap P = F_j$, then as we saw in the proof of Theorem 4, F_j can be written in the form

$$F_j = H_j \cap \bigcap_{i \neq j} H_i^+ = \bigcap_{i \neq j} (H_j \cap H_i^+),$$

a polyhedral set in H_j. Hence any facet of F_j is of the form

$$F = F_j \cap H_i = P \cap H_j \cap H_i = F_j \cap F_i.$$

That is, bearing in mind Theorem 6:

Lemma 1. <u>Any $(d-2)$-face of a d-polytope P is the intersection of two facets of P.</u>

From Lemma 1 we deduce immediately:

Theorem 9. <u>Let F^j be any proper j-face of a d-polytope P, and let $j \leq k < d$. Then F^j is the intersection of the k-faces of P which contain F^j.</u>

Proof. By the corollary to Theorem 6, there are $(j+1)$-, ... $(d-1)$-faces of P, say $F^{j+1}, \ldots, F^{d-1}$, such that

$$F^j \subset F^{j+1} \subset \ldots \subset F^{d-1}.$$

So, let F^{k+1} be any $(k+1)$-face of P containing F^j (with $F^{k+1} = P$ if $k = d-1$). By Lemma 1, each $(k-1)$-face of F^{k+1} is the intersection of facets (k-faces) of F^{k+1}, and then each $(k-2)$-face is the intersection of $(k-1)$-faces of F^{k+1}, and so on. We conclude, by induction, that F^j itself is the intersection of k-faces of F^{k+1}, and so of k-faces of P, which completes the proof of the theorem.

The following result will be used in Chapter 4.

Theorem 10. Let F be a j-face of a d-polytope P. Then there exists a $(d-j-1)$-face F' of P, such that $\dim \text{conv}(F \cup F') = d$.

Notice that by Theorem 2 of Chapter 1, this necessarily implies that $F \cap F' = \emptyset$.

Proof. If F is a facet of P, then F' can be any vertex of P not in F. The general result follows by induction from this. For, if $\dim F = j \leq d-2$, then F is properly contained in some facet G of P, and by the induction hypothesis, there is some $(d-j-2)$-face G' of G such that $\dim \text{conv}(F \cup G') = d-1$. Now let F' be any $(d-j-1)$-face of P containing G' but not contained in G. There is such a face F'; for clearly G' is contained in some facet of P apart from G, from which the existence of F' may be deduced by an easy induction argument. Then F' is the required face of P, since

$$\dim \text{conv}(F \cup F') > \dim \text{conv}(F \cup G') = d-1.$$

This completes the proof of the theorem.

The following characterization of a face of a polytope will be required in the next chapter.

Theorem 11. Let P be a convex polytope, and let $W \subseteq V = \text{vert } P$. Then $\text{conv } W$ is a face of P if and only if

$$\text{aff } W \cap \text{conv}(V \setminus W) = \emptyset.$$

Proof. If $F = \text{conv } W$ is a face of P, then there is a supporting hyperplane H of P such that $H \cap P = F$. Then $V \setminus W$ is contained in one of the open half-spaces bounded by H, so that $H \cap \text{conv}(V \setminus W) = \emptyset$. But aff $W = $ aff $F \subseteq H$, so that aff $W \cap \text{conv}(V \setminus W) = \emptyset$. This shows that the condition stated in the theorem is necessary.

To show that it is also sufficient, suppose $x \in $ aff W, $y \in \text{conv}(V \setminus W)$. It is clear that for a fixed point y, the infimum of $\|x - y\|$ is attained when the line aff $\{x, y\}$ is perpendicular to aff W. Since $\text{conv}(V \setminus W)$ is closed and bounded, the infimum of $\|x - y\|$ is also attained when y varies in $\text{conv}(V \setminus W)$, and then $y = x\Phi$ (where Φ is the nearest point map for $\text{conv}(V \setminus W)$). The hyperplane H' through $x\Phi$ and normal to $x - x\Phi$ supports $\text{conv}(V \setminus W)$ (Lemma 2 of §1.4), and so the hyperplane H through x parallel to H' is such that

$$H \cap \text{conv}(V \setminus W) = \emptyset.$$

But by the previous remark, since H is also perpendicular to aff $\{x, x\Phi\}$, it follows that $H \supseteq $ aff W. In other words, $H \cap P = \text{conv } W$, so that conv W is a face of P. This proves the theorem.

If P is a d-polytope, then the corollary to Theorem 6 clearly implies that P has faces of each dimension $j = 0, \ldots, d-1$. Let $\mathscr{F}(P)$ denote the set of all the faces of P, including the improper faces $\emptyset$ and P. Then one can show that $\mathscr{F}(P)$, partially ordered by inclusion, is a finite lattice, called the face-lattice of P.

To see this let us denote the lattice operations by $\wedge$ and $\vee$. Then Theorem 8 shows that $F_1 \cap F_2$ serves as a suitable definition for $F_1 \wedge F_2$. We must also show that there is a minimal face $F_1 \vee F_2$, containing the faces F_1 and F_2, such that $(F_1 \vee F_2) \vee F_3 = F_1 \vee (F_2 \vee F_3)$; clearly we cannot define $F_1 \vee F_2$ as $F_1 \cup F_2$ for this is, in general, not a face of P. The existence of a face $F_1 \vee F_2$ will follow from Theorem 13 of §2.2, and then we will have shown that $\mathscr{F}(P)$ is a lattice with $\emptyset$ and P as minimal and maximal elements, respectively. The concept of the face lattice leads to the following important idea:

Definition. Two polytopes P_1 and P_2 are said to be <u>combinatorially equivalent</u>, denoted by $P_1 \approx P_2$, if their face-lattices $\mathscr{F}(P_1)$ and $\mathscr{F}(P_2)$ are isomorphic.

In other words, P_1 and P_2 are combinatorially equivalent if there exists a one-to-one correspondence Ψ between the set of faces of P_1 and the set of faces of P_2 which preserves inclusion; that is, for every two faces F_1 and F_2 of P_1, $F_1 \subseteq F_2$ if and only if $F_1 \Psi \subseteq F_2 \Psi$.

Intuitively, P_1 and P_2 are combinatorially equivalent if one can be formed from the other by a 'distortion' which may change the shapes of the faces, but does not alter their number or arrangement. In particular, if Γ is an affinity, then the polytopes P and $P\Gamma$ are combinatorially equivalent, and a similar statement applies to non-singular permissible projective transformations.

The combinatorial theory of polytopes is concerned with the properties of the equivalence classes under combinatorial equivalence, rather than with polytopes themselves. On the other hand, to establish these properties, it will usually be necessary to choose one particular polytope from the class under consideration, and apply metrical or affine methods, rather than apply combinatorial arguments to the class itself.

We end this section with a result that may be regarded as a partial converse to Corollary 2 of Theorem 4. For this we make use of a special (n - 1)-simplex

$$T^{n-1} = \text{conv}\{e_1, \ldots, e_n\} \subset E^n,$$

where $e_i = (\delta_{i1}, \ldots, \delta_{in})$, and δ_{ij} is the Kronecker delta. The points $e_1, \ldots, e_n$ comprise a linear basis of E^n, and so are affinely independent. T^{n-1} is called regular, since all its edges are equal in length, and it may be written

$$T^{n-1} = \{(\xi_1, \ldots, \xi_n) \in E^n \mid \sum_{i=1}^{n} \xi_i = 1, \; \xi_i \geq 0, \; i = 1, \ldots, n\}.$$

T^{n-1} has exactly n facets, the points of the i-th facet being characterized, in the above representation, by the vanishing of ξ_i. Equivalently, if we write

$$H_i = \{(\xi_1, \ldots, \xi_n) \in E^n \mid \xi_i = 0\},$$

then H_i is a coordinate hyperplane which supports T^{n-1}, and the i-th facet of T^{n-1} is $H_i \cap T^{n-1}$. For further properties of simplices, see §2.3 (i).

Theorem 12. <u>Let P be a d-polytope (in E^d) with n facets, and let a $\in$ int P. Let T^{n-1} be the regular (n - 1)-simplex defined above, and let b $\in$ relint T^{n-1}. Then there exists a d-dimensional affine subspace $L \subset E^n$, containing b, such that</u>

(i)   <u>P is projectively equivalent to the d-polytope $L \cap T^{n-1}$, and</u>

58

(ii) *In this projective equivalence, a corresponds to b.*

Proof. Let us choose the coordinate system in E^d so that the point a coincides with the origin o. By Corollary 1 of Theorem 4, P is the intersection of exactly n closed half-spaces of E^d, one corresponding to each of its facets. Since $o \in \text{int } P$, we may therefore write

$$P = \{x \in E^d | \ \langle x, w_i \rangle \leq 1\},$$

where the w_i are the outward normals to the half-spaces, and, since P is bounded, $\{w_1, \ldots, w_n\}$ positively spans E^d.

Let

$$b = (\beta_1, \ldots, \beta_n) \in \text{relint } T^{n-1},$$

so that by Chapter 1, Theorem 6,

$$\sum_{i=1}^{n} \beta_i = 1, \quad \beta_i > 0,$$

and consider the transformation $\Theta: E^d \to E^n$ defined by

$$x\Theta = \frac{(\beta_1(1 - \langle x, w_1 \rangle), \ldots, \beta_n(1 - \langle x, w_n \rangle))}{\sum_{i=1}^{n} \beta_i(1 - \langle x, w_i \rangle)} \quad (3)$$

We define $L = E^d \Theta$, and claim that L is the required d-dimensional affine subspace.

To begin with, we see that Θ is a projective transformation. If Θ were singular, so that the rows of the corresponding matrix would be linearly dependent, then there exists a vector $(y, \gamma) = (\eta_1, \ldots, \eta_d, \gamma)$ such that $\langle w_i, y \rangle = \gamma$ for $i = 1, \ldots, n$. However, this is impossible, since the vectors w_i cannot all lie in a hyperplane. Hence Θ is non-singular, and $L = E^d \Theta$ is a d-dimensional affine subspace of E^n.

It is clear that $o\Theta = b$, and $P\Theta \subseteq T^{n-1}$, since the coordinates on the right of (3) are non-negative, and sum to 1. Finally we note that, in fact, $P\Theta = L \cap T^{n-1}$, for the ith facet of P is contained in the hyperplane

$$\{x \in E^d \mid \langle x, w_i \rangle = 1\},$$

which is mapped by Θ into the supporting hyperplane $H_i \cap L$ of $L \cap T^{n-1}$. This completes the proof of Theorem 12.

The map Θ gives a one-to-one correspondence between the set of facets of P and the set of facets of T^{n-1}. Clearly this correspondence may be chosen in many ways, but it can be proved that, once a particular choice of the correspondence has been made, the map Θ satisfying the conditions of Theorem 12 is unique.

Because of Theorem 8 of Chapter 1, a trivial generalization is that every d-polytope with n facets is projectively equivalent to a suitable d-dimensional section of any $(n-1)$-simplex. However in our application of Theorem 12 in Chapter 3 we shall only require the following weaker form: Every d-polytope with n facets is combinatorially equivalent to a d-dimensional section of a regular $(n-1)$-simplex $T^{n-1} = \text{conv}\{x_1, \ldots, x_n\}$ passing through the centroid $\frac{1}{n} \sum_{i=1}^{n} x_i$ of T^{n-1}.

2.2 POLARITY AND DUALITY

Definition. Let P be a polytope. Then a polytope P* is said to be <u>dual</u> to P if there is a one-to-one correspondence between the set of faces of P and the set of faces of P* which reverses the relation of inclusion.

Equivalently, P* is a dual of P if and only if the corresponding face lattices $\mathscr{F}$(P) and $\mathscr{F}$(P*) are anti-isomorphic. Consequently any two duals of a polytope P are combinatorially equivalent, and any dual (P*)* = P** of the dual P* is combinatorially equivalent to P. Thus a dual P* of P is a representative of a combinatorial equivalence class of polytopes, and is not a uniquely defined polytope. However, it is clear from the definition that, whichever particular representative P* is chosen, dim P* = dim P.

As an example, consider the triangular prism and triangular bipyramid in E^3, illustrated in Figure 15. (The terms 'prism' and 'bipyramid' will be defined generally in §2.3 (iv) and (iii).) Then the correspondence in which the faces F_i^0, F_j^1, F_k^2 of the prism are mapped into the faces G_i^2, G_j^1, G_k^0 of the pyramid, respectively, clearly reverses the relation of inclusion, so that the two polytopes are dual. (For clarity, the faces 'at the back of' each polytope have not been labelled in the diagram.)

The next theorem (Theorem 13) will establish the fact that every polytope possesses a dual. To prove this, we use the concept of a polar set.

Definition. Let K be a closed bounded convex set in E^d, with $o \in \text{int } K$. Then the <u>polar set</u> K* of K is defined by

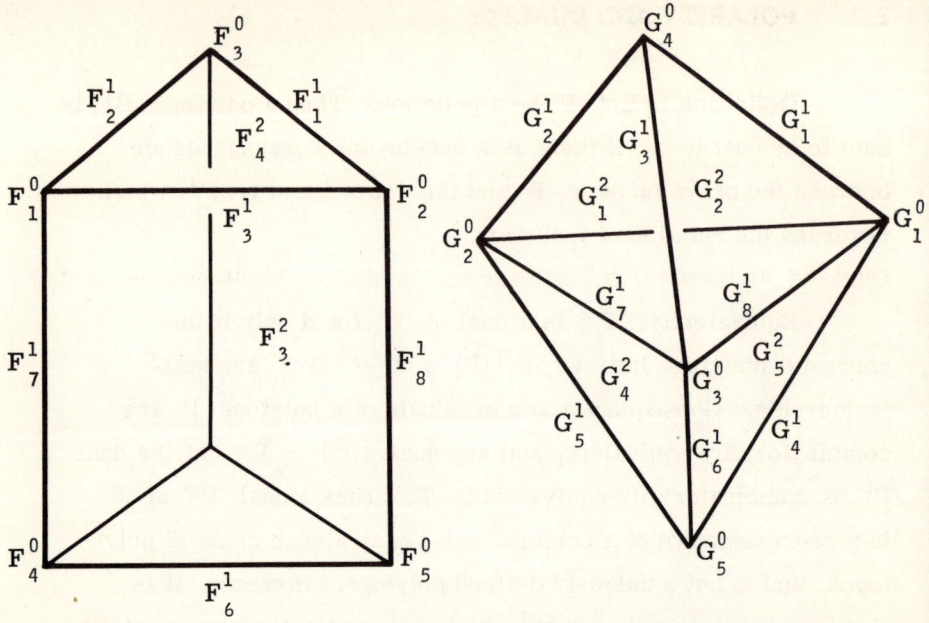

Figure 15

$$K^* = \{y \in E^d | \ \langle x, y \rangle \leq 1 \text{ for all } x \in K\} .$$

Theorem 13 will show that the polar set of a polytope P (with o ∈ int P) is a polytope dual to P. However, we first need a number of lemmas, to establish the basic properties of polar sets. Throughout these lemmas K, with or without subscripts, will represent a closed bounded convex set in E^d with o ∈ int K (so that K is d-dimensional), and K* will be its polar set.

Lemma 2. *K* is a closed set.*

Proof. It follows from the definition that

$$K^* = \bigcap_{x \in K} \{y \in E^d | \langle x, y \rangle \leq 1\}.$$

Thus K* is the intersection of a family of closed half-spaces, and so is itself closed.

Lemma 3. *If* $K_1 \subseteq K_2$, *then* $K_2^* \subseteq K_1^*$.

Proof. If $y \in K_2^*$, then by definition $\langle x, y \rangle \leq 1$ for all $x \in K_2$. Since $K_1 \subseteq K_2$, this implies that $\langle x, y \rangle \leq 1$ for all $x \in K_1$, and so $y \in K_1^*$. Hence, $K_2^* \subseteq K_1^*$.

Lemma 4. *K* is bounded, and* $o \in \text{int } K^*$.

Proof. Let $B(o, \rho)$ be the closed ball with centre o and radius $\rho > 0$. Clearly $(B(o, \rho))^* = B(o, \rho^{-1})$. Since $o \in \text{int } K$, there is a $\rho > 0$ such that $B(o, \rho) \subseteq K$. Hence by Lemma 3, $K^* \subseteq B(o, \rho^{-1})$, and therefore K* is bounded.

Similarly, since K is bounded, there is a $\sigma > 0$ such that $K \subseteq B(o, \sigma)$. By Lemma 3 again, $B(o, \sigma^{-1}) \subseteq K^*$, which shows that $o \in \text{int } K^*$.

Lemma 5. *K* is convex.*

Proof. As in the proof of Lemma 2, we can write

$$K^* = \bigcap_{x \in K} \{y \in E^d | \langle x, y \rangle \leq 1\}.$$

Thus K* is the intersection of a family of convex sets, and so is itself convex.

Lemma 6. $K^{**} = K$.

Proof. If $x \in K$, then $\langle x, y \rangle \leq 1$ for all $y \in K^*$, and so $x \in K^{**}$. Hence $K \subseteq K^{**}$.

Conversely, suppose that $x_0 \notin K$, and let $x_0 \Phi$ be the image of x_0 under the nearest point map Φ for K. Then Lemma 2 of Chapter 1 gives a construction for a supporting half-space H^- of K, say

$$H^- = \{x \in E^d | \langle x, y_0 \rangle \leq 1\},$$

whose boundary H contains $x\Phi$. (y_0 is a scalar multiple of $x_0 - x_0 \Phi$.) So, $\langle x, y_0 \rangle \leq 1$ for all $x \in K$, but $\langle x_0, y_0 \rangle > 1$. The first inequality implies that $y_0 \in K^*$, and so the second implies that $x_0 \notin K^{**}$. Thus $K^{**} \subseteq K$.

These two inclusions show that $K^{**} = K$, and the lemma is proved.

Now let F be any face of K, proper or improper, and let

$$\hat{F} = \{y \in K^* | \langle y, x \rangle = 1 \text{ for all } x \in F\}.$$

Lemma 7. $\hat{F}$ *is a face of* K^*.

Proof. Since $\hat{\emptyset} = K^*$, $\hat{K} = \emptyset$, we may suppose that F is a proper face of K. Let $x_0 \in \text{relint } F$, and let

$$F^* = \{y \in K^* | \langle y, x_0 \rangle = 1\}.$$

Then F^* is a face of K^*, and clearly $\hat{F} \subseteq F^*$.

Suppose that $y_0 \in K^* \setminus \hat{F}$. Then there is an $x_1 \in F$ such that $\langle y_0, x_1 \rangle < 1$. Since $x_0 \in$ relint F, there is an $x_2 \in F$ such that

$$x_0 = (1 - \lambda)x_1 + \lambda x_2$$

for some λ with $0 < \lambda < 1$. Since $y_0 \in K^*$, we have $\langle y_0, x_2 \rangle \leq 1$, and so

$$\langle y_0, x_0 \rangle = (1 - \lambda) \langle y_0, x_1 \rangle + \lambda \langle y_0, x_2 \rangle < 1.$$

Consequently $y_0 \notin F^*$. It follows that $F^* \subseteq \hat{F}$, and combining this with the previous inclusion, we have $F^* = \hat{F}$. Thus $\hat{F}$ is a face of K^*.

Lemma 8. <u>The map which sends each face F of K into the face $\hat{F}$ of K^* defined above is one-to-one and inclusion reversing.</u>

Proof. By an argument analogous to that used in the proof of Lemma 3, we see that the map is inclusion reversing. To prove that it is one-to-one it is enough to show that $\hat{\hat{F}} = F$. Now by definition

$$\hat{\hat{F}} = \{x \in K^{**} | \langle x, y \rangle = 1 \text{ for all } y \in \hat{F}\},$$

and so, since $K^{**} = K$ (Lemma 6), this shows that $F \subseteq \hat{\hat{F}}$.

The result is trivial if $F = \emptyset$ or $F = K$, so let F be a proper face of K. Then there is a supporting hyperplane of K,

$$H = \{x \in E^d | \langle x, y_0 \rangle = 1\},$$

such that $F = H \cap K$, and

$$K \subset H^- = \{x \in E^d | \ \langle x, y_0 \rangle \leq 1\},$$

which implies that $y_0 \in \hat{F}$. If $x_0 \in K\backslash F$, then $\langle x_0, y_0 \rangle < 1$, and so $x_0 \notin \hat{\hat{F}}$. Hence $\hat{\hat{F}} \subseteq F$, which together with the previous inclusion implies $\hat{\hat{F}} = F$. This proves Lemma 8.

We are now in a position to prove the important theorem:

Theorem 13. <u>Let P be a d-polytope in E^d, with $o \in \text{int } P$. Then the polar set P^* of P is a d-polytope dual to P.</u>

Proof. We have already found the one-to-one inclusion reversing correspondence between the faces of P and those of P^*. It is therefore sufficient to show that P^* is a polytope. Let vert $P = \{w_1, \ldots, w_n\}$, and write

$$Q = \{y \in E^d | \ \langle y, w_i \rangle \leq 1, i = 1, \ldots, n\}.$$

We shall show that $Q = P^*$.

Firstly, since vert $P \subset P$, it follows immediately that $P^* \subseteq Q$. Now let $y \in Q$, so that $\langle y, w_i \rangle \leq 1$ for $i = 1, \ldots, n$. Since by Theorem 2, $P = \text{conv}(\text{vert } P)$, a point $x \in P$ can be written in the form

$$x = \sum_{i=1}^{n} \lambda_i w_i, \quad \lambda_i \geq 0, \quad \sum_{i=1}^{n} \lambda_i = 1,$$

and so

$$\langle y, x \rangle = \langle y, \sum_{i=1}^{n} \lambda_i w_i \rangle = \sum_{i=1}^{n} \lambda_i \langle y, w_i \rangle \leq 1.$$

Since this is true for every point $x \in P$, by definition $y \in P^*$, and so $Q \subseteq P^*$.

We deduce that $P^* = Q$. Since P^* is bounded (by Lemma 4), and Q is a polyhedral set, Theorem 4 implies that P^* is a polytope, which completes the proof of the theorem.

We remark that with the definition $F_1 \vee F_2 = \widehat{(\hat{F}_1 \cap \hat{F}_2)}$, Theorem 13 completes the proof of the fact that the face-lattice $\mathscr{F}(P)$ of a polytope P is a lattice. ;

Duality is, of course, a combinatorial concept; the next result shows, however, that the transformations most naturally associated with polar sets are projective.

Theorem 14. <u>Let K be a closed bounded convex set in E^d with $o \in$ int K, and let Θ be a non-singular projective transformation of E^d, permissible for K, such that $o \in$ int $K\Theta$. Let K^*, $(K\Theta)^*$ be the polar sets of K, $K\Theta$ respectively. Then there is a (non-singular) projective transformation Θ^* (permissible for K^*), such that $K^*\Theta^* = (K\Theta)^*$.</u>

Proof. We first show that the transformation Θ takes the particular form

$$x\Theta = \frac{(x - p)\Lambda}{1 - \langle x, q \rangle} \qquad (4)$$

where $p \in$ int K, $q \in$ int K^*, and Λ is a non-singular linear transformation. For, let Θ be

$$x\Theta = \frac{x\Gamma}{\langle x, a \rangle + \beta}$$

where Γ is an affine transformation, and $(a, \beta) \neq 0$. Since $o \in \text{int}(K\Theta)$, there is a $p \in \text{int } K$ such that $p\Theta = o$. Then $p\Gamma = o$, so that

$$x\Theta = \frac{x\Gamma - p\Gamma}{\langle x, a \rangle + \beta}$$

$$= \frac{(x - p)\Lambda'}{\langle x, a \rangle + \beta},$$

for some (non-singular) linear transformation Λ'. Since $o \in \text{int } K$, and Θ is permissible for K, we must have $\beta \neq 0$, so dividing numerator and denominator by β we deduce

$$x\Theta = \frac{(x - p)\Lambda}{1 - \langle x, q \rangle},$$

where $\Lambda = \beta^{-1}\Lambda'$, and $q = -\beta^{-1}a$. Again, since Θ is permissible for K, $o \in K$ and $1 - \langle o, q \rangle = 1 > 0$, we must have $1 - \langle x, q \rangle > 0$ for all $x \in K$, and so (comparing the definition of a polar set), we conclude that $q \in \text{int } K^*$.

Having written Θ in this particular form, the remainder of the proof follows easily. From (4),

$$K\Theta = \{x' = \frac{(x - p)\Lambda}{1 - \langle x, q \rangle} \mid x \in K\},$$

hence for all $y \in (K\Theta)^*$, $x \in K$,

$$\left\langle y, \frac{(x - p)\Lambda}{1 - \langle x, q \rangle} \right\rangle \leq 1,$$

which, after a little manipulation, is seen to be equivalent to

$$\left\langle x, \frac{y\Lambda^* + q}{1 + \langle y\Lambda^*, p \rangle} \right\rangle \leq 1 ,$$

where Λ^* is the adjoint transformation of Λ, defined by

$$\langle u\Lambda^*, z \rangle = \langle u, z\Lambda \rangle ,$$

for all $u, z \in E^d$. Λ^* is a non-singular linear transformation. In other words,

$$K^* = \left\{ y' = \frac{y\Lambda^* + q}{1 + \langle y\Lambda^*, p \rangle} \;\middle|\; y \in (K\Theta)^* \right\} . \tag{5}$$

The transformation in (5) is a non-singular projective transformation, whose inverse is seen to be

$$y = \frac{y'(1 - \langle q, p \rangle)(\Lambda^*)^{-1}}{1 - \langle y', p \rangle} - q(\Lambda^*)^{-1} ,$$

or, changing the notation,

$$y\Theta^* = \frac{(y - q)M}{1 - \langle y, p \rangle} ,$$

where $M = N(\Lambda^*)^{-1}$, and N is the (non-singular) linear transformation

$$zN = z(1 - \langle q, p \rangle) + \langle z, p \rangle q .$$

Thus we have shown that

$$(K\Theta)^* = K^*\Theta^* ,$$

which proves the theorem.

In the case of convex polytopes, an immediate consequence is:

Corollary. <u>Let P_1 and P_2 be projectively equivalent polytopes in E^d with $o \in \text{int } P_1$, $o \in \text{int } P_2$. Then their polar sets P_1^* and P_2^* are also projectively equivalent.</u>

We have seen in Theorem 12 that any polytope is projectively equivalent to a section of some simplex. The next result connects sections and projections of polytopes.

Theorem 15. <u>Let P be a d-polytope in E^d with $o \in \text{int } P$, and let P^* be the polar set of P. Let L be an r-dimensional linear subspace of E^d through o, and let Π represent orthogonal projection onto L. Then $L \cap P^*$ is the polar set in L of $P\Pi$.</u>

Proof. Each vertex w_i ($i = 1, \ldots, n$) of P can be written in the form $w_i = w_i' + w_i''$, with $w_i' \in L$, and $w_i'' \in L^\perp$, the orthogonal complement in E^d of L. Then, as in the proof of Theorem 13, $P^* = \bigcap_{i=1}^{n} H_i^+$, where H_i^+ is the closed half-space

$$H_i^+ = \{y \in E^d \mid \langle y, w_i \rangle \leq 1\},$$

and $L \cap P^* = \bigcap_{i=1}^{n} (L \cap H_i^+)$. But $y \in L \cap H_i^+$ if and only if

$$\langle y, w_i \rangle = \langle y, w_i' + w_i'' \rangle =$$

$$\langle y, w_i' \rangle + \langle y, w_i'' \rangle \leq 1,$$

and $y \in L$, that is, $\langle y, w_i'' \rangle = 0$. Hence

$$L \cap H_i^+ = \{y \in L | \langle y, w_i' \rangle \leq 1\}.$$

Consequently $L \cap P^*$ is the polar set (in L) of the polytope

$$\text{conv}\{w_1', \ldots, w_n'\} = \text{conv}\{w_1 \Pi, \ldots, w_n \Pi\}$$
$$= (\text{conv}\{w_1, \ldots, w_n\})\Pi = P\Pi.$$

This proves the theorem.

We remark that if $w_i \Pi$ is not a vertex of $P\Pi$, then the corresponding half-space $L \cap H_i^+$ in L is redundant in defining the polar set $(P\Pi)^*$. Thus if every vertex of P projects into a vertex of $P\Pi$, then L will meet every facet of P^* in a relatively interior point.

Another application of the concept of duality between polytopes is the following.

Theorem 16. <u>Let F_1 be a j-face, and F_2 a k-face of a polytope P, such that $F_1 \subseteq F_2$. Then the sublattice of the face-lattice $\mathscr{F}(P)$ of P consisting of all faces F of P such that $F_1 \subseteq F \subseteq F_2$ is isomorphic to the face-lattice of some $(k-j-1)$-polytope, which we denote by F_2/F_1.</u>

Notice that, as with duals, F_2/F_1 is not a uniquely defined polytope, but is a class of combinatorially equivalent polytopes.

Proof. The situation may be represented by

$$\emptyset \subseteq F_1 \subseteq F_2 \subseteq P \ .$$

The k-face F_2 is a k-polytope, and by Theorem 5, F_1 is a j-face of F_2. Consider the dual F_2^* of F_2. Then the face $\hat{F}_1$ of F_2^* corresponding to F_1 is (k - j - 1)-dimensional, and we have

$$\emptyset = \hat{F}_2 \subseteq \hat{F}_1 \subseteq \hat{\emptyset} = F_2^* \ .$$

The lattice $\mathscr{F}(\hat{F}_1)$ of faces of $\hat{F}_1$ is clearly anti-isomorphic to the required sublattice, and so, taking the dual of $\hat{F}_1$, we obtain

$$F_2/F_1 \approx (\hat{F}_1)^* \ .$$

This proves Theorem 16.

The following two corollaries of the theorem are immediate consequences of the definition of the polytope F_2/F_1.

Corollary 1. <u>If $\hat{F}_1$, $\hat{F}_2$ are the faces of P^* corresponding to the faces F_1, F_2 of P, then</u>

$$(F_2/F_1)^* \approx \hat{F}_1/\hat{F}_2 \ .$$

Corollary 2. <u>If $F_1 \subseteq F_2 \subseteq F_3$ are faces of P, then F_2/F_1 is combinatorially equivalent to some face of F_3/F_1, and</u>

$$(F_3/F_1)/(F_2/F_1) \approx F_3/F_2 \ .$$

Definition. Let F be a vertex of a polytope P, and let H be a hyperplane strictly separating F from the remaining vertices vert $P \backslash F$ of P. Then $H \cap P$ is called a <u>vertex-figure</u> of P at F.

The idea of a vertex-figure occurs frequently in the classical literature on polytopes. A polytope has a vertex-figure at every one of its vertices (with reference to the proof of Theorem 2 in §2.1, it may be taken to be the hyperplane through the mid-point of, and normal to the line segment $\text{conv}\{x_i, x_i\Phi\}$, where $F = \{x_i\}$ and $x_i\Phi$ is the nearest point of $\text{conv}(\text{vert } P \backslash F)$ to x_i). Since H contains no vertices of P, it meets the relative interior of each face of P which contains F, and for each pair F_1, F_2 of faces of P such that $F \subseteq F_1 \subseteq F_2$, we have $H \cap F_1 \subseteq H \cap F_2$. We deduce the following result.

Theorem 17. <u>A vertex-figure $H \cap P$ of a polytope P, at a given vertex F of P, is a polytope combinatorially equivalent to P/F.</u>

This theorem leads to a direct proof of Theorem 16, without the use of duality. For if F_1 is a j-face of P, by the corollary to Theorem 6 there are 0-, 1-, ..., (j - 1)-faces $F^0, F^1, \ldots, F^{j-1}$ of P, such that

$$F^0 \subset F^1 \subset \ldots \subset F^{j-1} \subset F_1.$$

So, if F_2 is a k-face of P such that $F_1 \subseteq F_2$, for each $i = 0, \ldots, j$, F_2/F^i is combinatorially equivalent to the vertex-figure of F_2/F^{i-1} at its vertex F^i/F^{i-1} (we write $F^{-1} = \emptyset$, $F^j = F_1$). This gives an inductive construction for F_2/F_1.

2.3 SOME SPECIAL TYPES OF POLYTOPES

In this section we shall describe some particular examples of polytopes to which we shall refer in the sequel. These examples also serve to illustrate many of the concepts already introduced.

(i) <u>Simplices</u>

The simplest type of d-polytope is a d-simplex, the analogue of the triangle (in E^2) and tetrahedron (in E^3). We recall the definition from §1.1: If V is any set of d + 1 affinely independent points, then $T^d = \text{conv } V$ is called a d-simplex.

Every face of T^d is the convex hull of some subset of V, and since every subset of an affinely independent set is affinely independent, it is itself a simplex of some dimension ≤ d. Without loss of generality, suppose that $T^d \subset E^d$. Then any subset W of V containing exactly d points spans a hyperplane of E^d, and this hyperplane clearly supports T^d. Hence conv W is a (d - 1)-simplex, which is a facet of T^d. Since every face of conv W is a face of T^d (Theorem 7), an induction argument yields the following:

Proposition 1. <u>Every k-face</u> (0 ≤ d ≤ d - 1) <u>of a d-simplex</u> T^d <u>is a k-simplex, and every</u> k + 1 <u>vertices of</u> T^d <u>are the vertices of a k-face of</u> T^d. <u>The number of k-faces of</u> T^d <u>is</u>

$$f_k(T^d) = \binom{d+1}{k+1}.$$

From Proposition 1 we immediately deduce:

Proposition 2. *Any two d-simplices are combinatorially equivalent.*

This result also follows from Theorem 7 of Chapter 1.

A d-simplex T^d in E^d is the intersection of the $d+1$ closed half-spaces which contain T^d and are bounded by the hyperplanes containing the facets of T^d (Theorem 4, Corollary 1). Thus the polar set of T^d (with respect to an interior point) is a d-polytope with $d+1$ vertices (which must therefore be affinely independent). Hence:

Proposition 3. *A dual of a d-simplex is also a d-simplex.*

(ii) **Pyramids**

A d-simplex T^d is the convex hull of a $(d-1)$-simplex T^{d-1} (any one of its facets), and a point $x \notin \text{aff } T^{d-1}$ (the remaining vertex). We generalize this concept as follows.

A d-*pyramid* P is the convex hull of a $(d-1)$-polytope Q, called the *basis* of P, and a point $x \notin \text{aff } Q$, called the *apex* of P.

Let F be a k-face of P, determined by the hyperplane H, so that $F = H \cap P$. Since vert $F \subset$ vert $P =$ vert $Q \cup \{x\}$, there are two possibilities.

(a) $x \notin$ vert F. Then Theorem 5 implies that F is a k-face of Q. (Notice that F can be Q itself.)

(b) $x \in$ vert F. Then vert $F \setminus \{x\} \subset$ vert Q, and is the set of vertices of a $(k-1)$-face $H \cap Q = F \cap Q$ of Q (Theorem 8 and Theorem 5). This implies that F is the pyramid with basis $F \cap Q$ and apex x.

Conversely, by Theorem 7, every face of Q (including Q itself) is a proper face of P. If F is a proper face of Q, then there is a hyperplane H_0 in aff Q such that $F = H_0 \cap Q$. It is clear that $H = \text{aff}(H_0 \cup \{x\})$ is a supporting hyperplane of $P = \text{conv}(Q \cup \{x\})$, and that $H \cap P = \text{conv}(F \cup \{x\})$ is the pyramid with basis F and apex x. We also notice that $\{x\}$ is a proper face of P (the hyperplane through x parallel to aff Q meets P in x alone). These results may be summarized as follows:

Proposition 4. *If P is a d-dimensional pyramid with basis Q (so that dim $Q = d - 1$) then the numbers of faces of P are given by*

$$f_k(P) = f_k(Q) + f_{k-1}(Q),$$

where we adopt the conventions $f_{-1}(Q) = f_{d-1}(Q) = 1$, $f_k(Q) = 0$ *for* $k < -1$ *or* $k > d - 1$.

We shall maintain these conventions throughout. The above discussion also establishes:

Proposition 5. *If $Q_1 \approx Q_2$, and for $i = 1, 2$, P_i is a pyramid with basis Q_i, then $P_1 \approx P_2$.*

An r-fold pyramid is defined recursively as follows: Any d-polytope is a 0-fold d-pyramid and it is its own basis. An r-fold d-pyramid P is a pyramid whose basis is an $(r - 1)$-fold $(d - 1)$-pyramid Q, and the basis of P is the same as the basis of Q. Hence the basis of an r-fold d-pyramid has dimension $d - r$.

Proposition 6. _If P is an r-fold d-pyramid with basis Q_ (dim Q = d - r), _then for all_ k, _with the usual conventions,_

$$f_k(P) = \sum_{i=0}^{r} \binom{r}{i} f_{k-i}(Q).$$

Since the basis of a (d - 1)-fold d-pyramid is a 1-polytope, which is a 1-fold 1-pyramid, the following holds:

Proposition 7. _A_ (d - 1)-_fold d-pyramid is also a d-fold d-pyramid; it is a d-simplex._

(iii) **Bipyramids**

Let Q be a (d - 1)-polytope, and let I be a closed line segment (1-polytope), such that relint Q ∩ relint I is a single point. Then the d-polytope P = conv (Q ∪ I) is called a d-**bipyramid** with **basis** Q.

If I = conv $\{x_0, x_1\}$, then by similar reasoning to that for pyramids, we see that each face of P is one of the following:

(a)   a proper face of Q,
(b)   a pyramid with a proper face of Q as basis, and either x_0 or x_1 as apex or
(c)   one of the vertices $\{x_0\}$ or $\{x_1\}$.

We immediately deduce the following two results:

Proposition 8. _Let_ P _be a d-bipyramid with basis_ Q (dim Q = d - 1). _Then, with the usual conventions,_

$$f_k(P) = f_k(Q) + 2f_{k-1}(Q), \quad 0 \le k \le d - 2,$$

$$f_{d-1}(P) = 2f_{d-2}(Q).$$

Proposition 9. *If* $Q_1 \approx Q_2$, *and for* $i = 1, 2$, P_i *is a bipyramid on* Q_i, *then* $P_1 \approx P_2$.

In analogy to pyramids, we can define recursively r-fold bipyramids: a general d-polytope is a 0-fold d-bipyramid, and an r-fold d-bipyramid is a bipyramid whose basis is an (r - 1)-fold (d - 1)-bipyramid. The analogue of Proposition 7 is:

Proposition 10. A (d - 1)-fold d-bipyramid is also a d-fold d-bipyramid.

Such a polytope is (combinatorially equivalent to) a d-crosspolytope; it is the analogue of the octahedron in E^3. Generally, if $e_1, \ldots, e_d$ are d linearly independent vectors in E^d, then $X^d = \text{conv } \{\pm e_1, \ldots, \pm e_d\}$ is called a d-crosspolytope, and if $e_1, \ldots, e_d$ are mutually orthogonal and of equal length, then X^d is said to be regular. From this construction, or inductively by observing that X^d is a bipyramid with basis X^{d-1}, we obtain:

Proposition 11. *The numbers of faces of the d-crosspolytope* X^d *are given by the equations*

$$f_k(X^d) = 2^{k+1} \binom{d}{k+1}, \quad -1 \leq k \leq d - 1.$$

(iv) Prisms

Let Q be a (d - 1)-polytope in E^d, and let $I = \text{conv } \{o, x\}$ be a closed line segment such that aff I is not parallel to any line in aff Q. Then the vector sum $P = Q + I$, defined by

$$Q + I = \{y \in E^d | y = z + w, \; z \in Q, \; w \in I\},$$

is a d-polytope called a d-__prism__ with __basis__ Q. It is clear that P = conv (Q ∪ (Q + x)), the convex hull of Q and its translate Q + x.

Let F be a proper face of P, and let H be a hyperplane in E^d such that F = H ∩ P. From the remark above, it follows that vert P ⊆ vert Q ∪ vert (Q + x), and so there are three possibilities:

(a) H ∩ (Q + x) = ∅. Then H ∩ Q = F, so that F is a face of Q.

(b) H ∩ Q = ∅, in which case F is a face of Q + x.

(c) H ∩ Q ≠ ∅, H ∩ (Q + x) ≠ ∅. Then H is a supporting hyperplane of both Q and Q + x. Since Q + x is a translate of Q, if H ∩ Q = F_1 then H ∩ (Q + x) = F_1 + x, and so F is a prism with basis F_1.

Conversely, any face of Q (including Q itself) is a face of P; similarly for Q + x. If F is a proper face of Q, then the prism F + I is a face of P. This description of the faces of P leads immediately to the following:

Proposition 12. __Let P be a d-prism with basis Q.__ (dim Q = d - 1.) __Then, with the usual conventions,__

$f_0(P) = 2f_0(Q)$,

$f_k(P) = 2f_k(Q) + f_{k-1}(Q)$, $1 \leq k \leq d$.

Proposition 13. __If__ $Q_1 \approx Q_2$, __and for__ i = 1, 2, P_i __is a prism with basis__ Q_i, __then__ $P_1 \approx P_2$.

Again, in analogy to both pyramids and bipyramids, we can define r-fold prisms: any d-polytope is a 0-fold prism, and an r-fold d-prism is a prism whose basis is an (r - 1)-fold (d - 1)-prism. The analogue of Propositions 7 and 10 is:

Proposition 14. <u>A (d - 1)-fold d-prism is also a d-fold d-prism.</u>

Such a polytope is called a d-<u>parallelotope,</u> the simplest example of which is a d-<u>cube</u> C^d. C^d is the vector sum of d mutually orthogonal equal line segments, say conv $\{o, e_i\}$ (i = 1, ..., d). If these line segments have equal length, we can take $e_1, ..., e_d$ to be an orthonormal basis of E^d, and with respect to this basis

$$C^d = \{x = (\xi_1, ..., \xi_d) \in E^d | 0 \le \xi_i \le 1, \ i = 1, ..., d\}.$$

From this construction, or inductively by observing that C^d is a prism with basis C^{d-1}, we deduce:

Proposition 15. <u>The numbers of faces of the d-cube</u> C^d <u>are given by the equations</u>

$$f_k(C^d) = 2^{d-k} \binom{d}{k}, \quad 0 \le k \le d.$$

From the above description of the faces of bipyramids and prisms, we obtain the following duality property.

Proposition 16. <u>Let Q* be a polytope dual to Q. Then any bipyramid with basis Q is dual to any prism with basis Q*.</u>

In particular, the d-crosspolytope X^d and the d-cube C^d are duals of each other.

(v) Simplicial and Simple Polytopes

Let X be any set of points in E^d. Then X is said to be in general position if every subset of X with $d + 1$, or fewer, points is affinely independent. If P is a d-polytope in E^d, and vert P is in general position, then no hyperplane in E^d contains more than d vertices of P. Consequently, every facet of P is a (d - 1)-simplex, and so, by Theorem 6 and Proposition 1, every proper face of P is a simplex. In an intuitive way, we may regard such a polytope as 'general', other polytopes being 'singular', in the sense that there are fortuitous affine dependences between their vertices.

This motivates the following definition: a polytope P is said to be simplicial if all its proper faces are simplices. Examples of simplicial d-polytopes are:

(a) any d-simplex,

(b) any d-bipyramid whose basis is a simplicial (d - 1)-polytope, and in particular,

(c) the d-crosspolytope X^d.

A prism is never simplicial, except in trivial cases.

Notice that we do not demand that the vertices of a simplicial polytope are in general position: more than d vertices of P may lie in a hyperplane H, but then H cannot support P.

The family of simplicial polytopes is very important in polytope theory, just as simplicial complexes are important in algebraic topology. We shall be continually referring to them in the sequel.

In a dual sense, a polytope P may be regarded as 'general' if it is a bounded polyhedral set defined by hyperplanes 'in general position' in E^d (Theorem 4). Precisely, this means that for $r \le d$, no subset of r of the hyperplanes intersects in a subspace of dimension greater than d - r, and every d + 1 of the hyperplanes have empty intersection. Generalizing this concept, we way that a d-polytope P is <u>simple</u> if every vertex of P lies in exactly d facets of P; it follows that every vertex-figure of P (§2.1) is a simplex.

Clearly, the dual of a simplicial polytope is simple (and conversely), and so, from a combinatorial point of view there is no advantage in considering one type rather than the other. However, we prefer to deal with simplicial polytopes; one reason is that sets of points are usually a little easier to visualize intuitively than sets of hyperplanes.

(vi) Cyclic Polytopes

Although discovered only fairly recently, cyclic polytopes already play an important role in the combinatorial theory of polytopes.

Let M be the moment curve in E^d, defined parametrically by $x(\tau) = (\tau, \tau^2, \ldots, \tau^d)$ $(-\infty < \tau < \infty)$. <u>A cyclic polytope</u> C(v, d) is the convex hull of any $v \ge d + 1$ distinct points $x(\tau_i)$ (i = 1, ..., v) on M. The reason we use the moment curve M in this definition is that it is the simplest example of a d-th order curve in E^d, that is to say, no hyperplane in E^d meets M in more than d points. We could equally well use any other d-th order curve (and many other authors have done so when writing about cyclic polytopes), but with the moment curve the calculations tend to become somewhat easier.

Let

$$V = \{x(\tau_i) | 1 \leq i \leq v, \quad \tau_1 < \tau_2 < \ldots < \tau_v\}$$

be the given subset of M. Our main purpose is to describe which subsets W of V are the vertices of faces of $C(v, d) = \text{conv } V$. Such subsets will be completely characterized in Proposition 18, and this will enable us to determine the numbers $f_k(v, d) = f_k(C(v, d))$ of faces of $C(v, d)$ of each dimension $1 \leq k \leq d - 1$. To begin with, we need some preliminary results.

Proposition 17. $C(v, d)$ is a simplicial polytope.

Proof. In fact we show that the vertices of $C(v, d)$ are in general position. Let W be any subset of V with card $W = d + 1$, and suppose, that the points of W correspond to the (distinct) parameters $\tau = \sigma_0, \ldots, \sigma_d$. From the criterion stated in §1.1, we see that W is affinely independent, since

$$\begin{vmatrix} 1 & \sigma_0 & \sigma_0^2 & \ldots & \sigma_0^d \\ 1 & \sigma_1 & \sigma_1^2 & \ldots & \sigma_1^d \\ \cdot & \cdot & \cdot & \ldots & \cdot \\ \cdot & \cdot & \cdot & \ldots & \cdot \\ \cdot & \cdot & \cdot & \ldots & \cdot \\ 1 & \sigma_d & \sigma_d^2 & \ldots & \sigma_d^d \end{vmatrix} = \prod_{0 \leq i < j \leq d} (\sigma_j - \sigma_i) \neq 0.$$

Hence no hyperplane of E^d can contain an affinely dependent subset of V, and so every face of $C(v, d)$ is a simplex. This proves the proposition.

If we write $x_i = x(\tau_i)$ ($i = 1, \ldots, v$), and $x_i < x_j$ if and only if $\tau_i < \tau_j$, then we see that $V = \{x_1, \ldots, x_v\}$ is a totally ordered set. For brevity, we shall call any such totally ordered set V, with card $V = v$, a v-<u>set</u>. If $W \subset V$, a subset $X \subseteq W$ will be called <u>contiguous</u> if for some $1 < i \leq j < v$

$$X = \{x_i, x_{i+1}, \ldots, x_j\}, \quad x_{i-1} \notin W, \quad x_{j+1} \notin W.$$

X will be called even or odd according to the parity of card $X = j - i + 1$. An <u>end-set</u> is a subset $Y \subseteq W$ of the form

$$Y = \{x_1, \ldots, x_i\}, \quad x_{i+1} \notin W, \quad \text{or}$$
$$Y = \{x_j, \ldots, x_v\}, \quad x_{j-1} \notin W.$$

Clearly any proper subset $W \subset V$ can be written uniquely in the form

$$W = Y_1 \cup X_1 \cup \ldots \cup X_t \cup Y_2,$$

where $0 \leq t \leq [\frac{1}{2}(v-1)]$, the X_i are contiguous subsets, and Y_1, Y_2 are end-sets of V or empty. W is said to be of type (r, s) if card $W = r$, and exactly s of the contiguous subsets X_i are odd. We can now state the main result:

Proposition 18. <u>Let</u> W <u>be any subset of</u> $V = $ vert $C(v, d)$ ($v \geq d + 1$). <u>Then</u> conv W <u>is a k-face of</u> $C(v, d)$ ($0 \leq k \leq d - 1$) <u>if and only if</u> W <u>is of type</u> $(k + 1, s)$, <u>for some</u> $0 \leq s \leq d - k - 1$.

Proof. By Proposition 17, $C(v, d)$ is simplicial, so if conv W is a k-face of $C(v, d)$, then card $W = k + 1$. Consider first the case $k = d - 1$. Given any subset $W \subset V$ with card $W = d$, then by Proposition 17, W is an affinely independent set, and so $H = \text{aff } W$ is a hyperplane in E^d. Since M is a d-th order curve, $H \cap M = W$, and the points of W divide M into $d + 1$ arcs lying alternately on each side of H. Now conv W is a facet of $C(v, d)$ if and only if H supports $C(v, d)$; that is, if and only if all the points of $V \setminus W$ lie on the same side of H. Clearly this will happen if and only if every two points of $V \setminus W$ are separated on M by an even number of points of W, and this, in turn, is equivalent to the condition that W is of type $(d, 0)$, that is, contains no odd contiguous subsets. The proposition is therefore true for $k = d - 1$. (The condition in this particular case is usually known as <u>Gale's evenness condition</u>.)

Consider now the general case. Let $W \subset V$, with card $W = k + 1$ ($0 \le k \le d - 1$) be the given subset. If W has at most $d - k - 1$ odd contiguous subsets, then it is clearly possible to find a subset $T \subset M$, with $T \cap V = \emptyset$ and card $T = d - k - 1$, such that the subset $T \cup W$ of the $(v + d - k - 1)$-set $T \cup V$ has only even contiguous subsets. Then the hyperplane $H = \text{aff } (T \cup W)$ supports the cyclic polytope $C(v + d - k - 1, d) = \text{conv } (T \cup V)$, and since $C(v, d) \subseteq \text{conv } (T \cup V)$, and $H \cap V = W$, by Theorem 5, $H \cap C(v, d) = \text{conv } W$ is a face of $C(v, d)$. The condition is also necessary, for by Theorem 6, if conv W is a face of $C(v, d)$, then it is also a face of some facet conv W' ($W \subseteq W' \subset V$) of $C(v, d)$. Since W' has no odd contiguous subsets, clearly W can have no more than $d - k - 1$ odd contiguous subsets. This completes the proof of the proposition.

Definition. A polytope P is said to be k-<u>neighbourly</u> if every subset of k points of $V = \text{vert } P$ is the set of vertices of a proper face of P.

The properties of neighbourly polytopes will be investigated in the next subsection, when the importance of the following corollary to Proposition 18 will become clear.

Corollary 1. $C(v, d)$ is $[\frac{1}{2}d]$-neighbourly.

Proof. If W is a subset of V such that card $W = [\frac{1}{2}d]$, then W can have at most $[\frac{1}{2}d]$ odd contiguous subsets. Since $[\frac{1}{2}d] \leq d - [\frac{1}{2}d]$, the proposition implies that conv W is a face of $C(v, d)$.

Corollary 2. The combinatorial type of a cyclic polytope $C(v, d)$ depends only upon v and d; it does not depend upon the choice of the subset $V \subset M$, or upon the particular choice of the d-th order curve M.

We shall now calculate the numbers of faces of $C(v, d)$ of various dimensions.

Proposition 19. The number $f_k(v, d)$ of k-faces of $C(v, d)$ is given by the expressions:

$$f_k(v, 2n) = \sum_{j=1}^{n} \frac{v}{v-j} \binom{v-j}{j} \binom{j}{k+1-j},$$

$$0 \leq k \leq 2n - 1, \qquad (1)$$

$$f_k(v, 2n+1) = \sum_{j=0}^{n} \frac{k+2}{v-j} \binom{v-j}{j+1} \binom{j+1}{k+1-j}$$

$$0 \leq k \leq 2n, \qquad (2)$$

according to whether the dimension d is even or odd.

Proof. The proof of the proposition depends upon a simple combinatorial argument to determine the number of distinct subsets $W \subset V$ of type $(k + 1, s)$, with $s \leq d - k - 1$. The even and odd dimensional cases are essentially different.

We begin with the case $d = 2n$. We shall require some transformations of the set W, and for this purpose we introduce the concept of a v-<u>circuit</u>. This is a set of v distinct points on an oriented simple closed curve N. Thus every point has a unique successor, and the v-th successor of every point is itself. Contiguous subsets of a v-circuit are defined in the obvious manner, and, as before, a subset W of a v-circuit is said to be of type (r, s) if card $W = r$, and W contains exactly s odd contiguous subsets.

Let V be a v-set, and let $W \subset V$ be a given subset of type $(k + 1, s)$ or $(k + 1, s - 1)$, where s is an integer satisfying $s \equiv k + 1 \pmod 2$. Then V may be made into a v-circuit V_1 by joining ends of an arc of M containing V (so that x_j is the successor of x_i in V_1 if and only if $j \equiv i + 1 \pmod v$), and then W becomes a subset $W_1 \subset V_1$ of type $(k + 1, s)$. (If W is of type $(k + 1, s - 1)$, then the condition $s \equiv k + 1 \pmod 2$ implies that the union of the end-sets of W has odd cardinality. Hence W_1 has one more odd contiguous subset than W, and so is of type $(k + 1, s)$.) Let us write $p(v, k + 1, s)$ for the total number of distinct subsets W_1 of V_1 of type $(k + 1, s)$, where $s \equiv k + 1 \pmod 2$ (which is the same as the number of subsets W of V of type $(k + 1, s)$ or $(k + 1, s - 1)$).

To determine the actual numerical value of $p(v, k + 1, s)$ we proceed as follows. Let W_2 be the subset of V_1 of type $(k + s + 1, 0)$ formed by adjoining to W_1 the successor of each of the s odd contiguous subsets of W_1. Then, writing $2j = k + s + 1$,

87

W_2 consists of j pairs of adjacent points of V_1. To each set W_2 clearly corresponds $\binom{j}{s}$ different sets W_1 (obtained by removing the second point of any s of the j pairs in W_2), and since the number of sets W_2 is $p(v, 2j, 0)$, we have

$$p(v, k+1, s) = \binom{j}{s} p(v, 2j, 0), \quad 2j = k + s + 1. \qquad (3)$$

So, we have reduced the problem to that of determining $p(v, 2j, 0)$. We delete one point of each pair in W_2, to obtain a subset W_3 of a $(v-j)$-circuit V_2 with card $W_3 = j$. The number of such subsets W_3 is clearly $\binom{v-j}{j}$. The relationship between corresponding sets W_2 and W_3 is found as follows. For each set W_3 let r be the number of cyclic permutations of V_2 (that is, automorphisms of V_2 which preserve order and orientation of N) which leave the subset W_3 invariant. Then r is also the number of cyclic permutations of V_1 which leave W_2 invariant. So, cyclic permutations of V_2 applied to W_3 yield $(v-j)/r$ distinct subsets of V_2, and cyclic permutations of V_1 applied to W_2 yield v/r distinct subsets of V_1 of type $(2j, 0)$. Hence

$$p(v, 2j, 0) = \frac{v}{v-j} \binom{v-j}{j} . \qquad (4)$$

So, from (3) and (4),

$$p(v, k+1, s) = \frac{v}{v-j} \binom{v-j}{j} \binom{j}{s} \quad (2j = k+s+1), \qquad (5)$$

and from Proposition 18,

$$f_k(v, 2n) = \sum_{\substack{s=0 \\ s \equiv k+1 \,(\mathrm{mod}\ 2)}}^{2n-k-1} p(v, k+1, s). \qquad (6)$$

Substituting the value of $p(v, k+1, s)$ from (5) in (6), and changing the variable in the summation from s to j, we immediately obtain (1). (Note that the terms corresponding to $1 \leq j < [\frac{1}{2}(k+1)]$ in the summations are identically zero; conventionally we put $\binom{p}{q} = 0$ if $q < 0$ or $q > p$.)

For the odd dimensional case $d = 2n+1$, we need to determine, for each k such that $0 \leq k \leq 2n$, the number of distinct subsets of $V \subset M$ of type $(k+1, s)$, with $s \leq 2n - k$. We make V into a $(v+1)$-circuit V_1 by joining the ends of an arc of M containing V, and adjoining one extra point x which is the successor of x_v and the predecessor of x_1. If W is a given subset of V, we define W_1 to be the corresponding subset of V_1, together with the additional point x. If W is of type $(k+1, s-1)$ or $(k+1, s)$, with $s \equiv k \pmod{2}$, then by similar reasoning to that for even dimension, W_1 is of type $(k+2, s)$. Using (6), the number of such subsets W_1 with $s \leq 2n - k$, is

$$\sum_{\substack{s=0 \\ s \equiv k \pmod{2}}}^{2n-k} p(v+1, k+2, s) = f_{k+1}(v+1, 2n+2). \quad (7)$$

For each subset W_1 of type $(k+2, s)$ ($s \equiv k \pmod{2}$), let r be the number of cyclic permutations of V_1 which leave W_1 invariant. Then cyclic permutations of V_1 applied to W_1 yield $(v+1)/r$ distinct subsets of V_1 of type $(k+2, s)$. Since deletion of any one of the $k+2$ points of W_1 converts V_1 into a v-set V, we see that each such set W_1 yields $(k+2)/r$ distinct subsets W of V of type $(k+1, s-1)$ or $(k+1, s)$. Hence, from (7), the total number of distinct subsets of V of type $(k+1, s)$, with $s \leq 2n - k$, is

$$f_k(v,\ 2n+1) = \frac{k+2}{v+1}\ f_{k+1}(v+1,\ 2n+2)\ .$$

If we substitute for $f_{k+1}(v+1,\ 2n+2)$ from (1), and change the variable in the summation from j to j - 1, we obtain (2), which completes the proof of the proposition.

It is of interest to explain why we carried out the calculations above in this particular way. As we remarked earlier, any d-th order curve can be used to construct the cyclic polytopes $C(v,\ d)$; in even dimension $d = 2n$, there are d-th order curves which are closed, an example of which is that parametrized by

$$x(\tau) = (\cos\tau,\ \sin\tau,\ \cos 2\tau,\ \sin 2\tau,\ \ldots,\ \cos n\tau,\ \sin n\tau)\ ,$$

$$0 \le \tau < 2\pi\ ,$$

which has also been used by various authors. For the odd dimensional case, we may use Gale's evenness condition to show that every vertex-figure of $C(v+1,\ 2n+2)$ is combinatorially equivalent to $C(v,\ 2n+1)$; this yields the relation (7), by means similar to those employed in §2.4 below.

The above method of determining the numbers of faces of the cyclic polytopes was first given in [9]. An alternative method will be indicated briefly in §2.4.

(vii) Neighbourly Polytopes

We have already remarked in (vi) that the cyclic polytope $C(v,\ d)$ is $[\frac{1}{2}d]$-neighbourly. Here we shall obtain some results about neighbourly polytopes in general.

Proposition 20. *If P is a k-neighbourly polytope, then every k vertices of P are affinely independent.*

Proof. Suppose that $W = \{w_1, \ldots, w_k\}$ is an affinely dependent subset of $V = \text{vert } P$. Without loss of generality, assume that $w_k \in \text{aff } \{w_1, \ldots, w_{k-1}\}$. Since $W \subset V$ we can find a point $z \in V \setminus W$. Let

$$Z = \{w_1, \ldots, w_{k-1}, z\}.$$

Since P is k-neighbourly, $F = \text{conv } Z$ is a face of P, and $Z = \text{vert } F$. Now for each hyperplane H such that $F = H \cap P$, we have

$$w_k \in \text{aff } \{w_1, \ldots, w_{k-1}\} \subseteq \text{aff } Z \subseteq H.$$

Hence $w_k \in F$ also. Since w_k is a vertex of P, this implies that $w_k \in \text{vert } F$. But $\text{vert } F = Z$, and $w_k \notin Z$, which is a contradiction. In other words, each subset W of V containing k points is affinely independent, which establishes the proposition.

From Proposition 20 we conclude that every k vertices of P determine a $(k-1)$-face of P, which is a $(k-1)$-simplex. Conversely, since every $(k-1)$-face of P contains at least k vertices, it follows that every $(k-1)$-face of P is a $(k-1)$-simplex. This has several consequences.

Proposition 21. *If P is a k-neighbourly polytope, and $1 \leq j \leq k$, then P is also j-neighbourly.*

This leads at once to:

Proposition 22. *If* P *is a k-neighbourly polytope with* v(> k) *vertices, then for each* $0 \leq j \leq k-1$,

$$f_j(P) = \binom{v}{j+1} .$$

Together with Theorem 5, Proposition 20 implies:

Proposition 23. *If* P *is a k-neighbourly polytope, and* W ⊆ vert P *is such that* card W > k, *then* conv W *is also a k-neighbourly polytope.*

Proposition 24. *Let* P *be a k-neighbourly d-polytope, with* $k > [\frac{1}{2}d]$. *Then* P *is a simplex.*

Proof. If P is not a d-simplex, then card (vert P) ≥ d + 2. Let W be any subset of vert P with card W = d + 2. Then by Radon's Theorem (Chapter 1, Theorem 9), W can be expressed as the union of two disjoint subsets Y and Z such that conv Y ∩ conv Z ≠ ∅. Without loss of generality we may assume card Y ≤ $[\frac{1}{2}(d+2)] = [\frac{1}{2}d] + 1 \leq k$. Since P is k-neighbourly, by Proposition 21 this implies that conv Y is a face of P. Then for every supporting hyperplane H of P such that H ∩ P = conv Y, we have H ∩ conv Z ≠ ∅, and hence H ∩ Z ≠ ∅. But this is impossible, since H contains no vertices of P other than the set Y.

This contradiction arose from the assumption that card (vert P) ≥ d + 2. We thus see that P is a simplex, which proves the proposition.

This proposition shows that, with the sole exception of the simplex, no d-polytope can be 'more neighbourly' than the cyclic polytopes $C(v, d)$ $(v \geq d + 2)$. About those polytopes with maximum neighbourliness, we have the following result:

Proposition 25. <u>Every n-neighbourly 2n-polytope P is simplicial.</u>

Proof. Let F be any facet of P. Then by Proposition 23, F is an n-neighbourly (2n - 1)-polytope, which by Proposition 24 is a simplex.

Since, however, any pyramid on a k-neighbourly polytope is also k-neighbourly, there are non-simplicial n-neighbourly (2n + 1)-polytopes. We quote without proof several other interesting results about neighbourly polytopes. If $v \leq 2n + 3$, then an n-neighbourly 2n-polytope with v vertices is combinatorially equivalent to $C(v, 2n)$, and a simplicial n-neighbourly (2n + 1)-polytope with v vertices is combinatorially equivalent to $C(v, 2n + 1)$. On the other hand the octahedron and the cyclic polytope $C(6, 3)$ are distinct simplicial 1-neighbourly 3-polytopes with 6 vertices, and an example is known of a 2-neighbourly 4-polytope with 8 vertices which is not combinatorially equivalent to $C(8, 4)$.

We conclude by remarking that there is no analogous theory for centrally symmetric polytopes (see [6]), but we do not have space to discuss this here.

2.4 EULER'S THEOREM AND THE DEHN-SOMMERVILLE EQUATIONS

Euler's Theorem was the first combinatorial theorem about polytopes to be discovered. To this day it remains one of the most important, and many generalizations and modifications of the original are known. The theorem is also a topological result about polytopes, but the following proof, which is one of many that exist, is elementary in nature, in that it uses no topological methods.

Theorem 18 (Euler's Theorem). Let P be a d-polytope, and let $f_j(P)$ denote the number of its j-faces ($0 \leq j \leq d-1$). Then

$$\sum_{j=0}^{d-1} (-1)^j f_j(P) = 1 + (-1)^{d-1} .$$

This equation is known as the Euler equation for the polytope P. It can also be written

$$\sum_{j=-1}^{d} (-1)^j f_j(P) = 0 ,$$

where we adopt the usual convention $f_{-1}(P) = f_d(P) = 1$, corresponding to the improper faces $\emptyset$ and P of P.

Proof. The theorem is obviously true for $d = 1$ ($f_0(P) = 2$) (and, in fact, for $d = 2$ also). We prove it generally by induction on the dimension d; let us assume, therefore, that it is true for polytopes of $d-1$ or fewer dimensions ($d \geq 2$).

Let P be a d-polytope in E^d, with $f_0(P) = v$ vertices. Let D be the union of at most $\frac{1}{2}v(v-1)$ hyperplanes through the origin, one perpendicular to each of the $\frac{1}{2}v(v-1)$ lines joining two vertices of P. Let $a \in E^d \setminus D$ be any vector, and let H be a hyperplane with normal a. Write $H_1, H_3, \ldots, H_{2v-1}$ for the v hyperplanes parallel to H, numbered in order, with one passing through each vertex of P. (By our choice of H, none of these hyperplanes can contain more than one vertex of P.) Let $H_2, \ldots, H_{2v-2}$ be $v - 1$ further hyperplanes parallel to H, such that, for $k = 1, \ldots, v - 1$, the hyperplane H_{2k} lies between H_{2k-1} and H_{2k+1}. Clearly H_1 and H_{2v-1} support P, and for each $i = 2, \ldots, 2v - 2$, $P_i = H_i \cap P$ is a $(d - 1)$-polytope. (See Figure 16.)

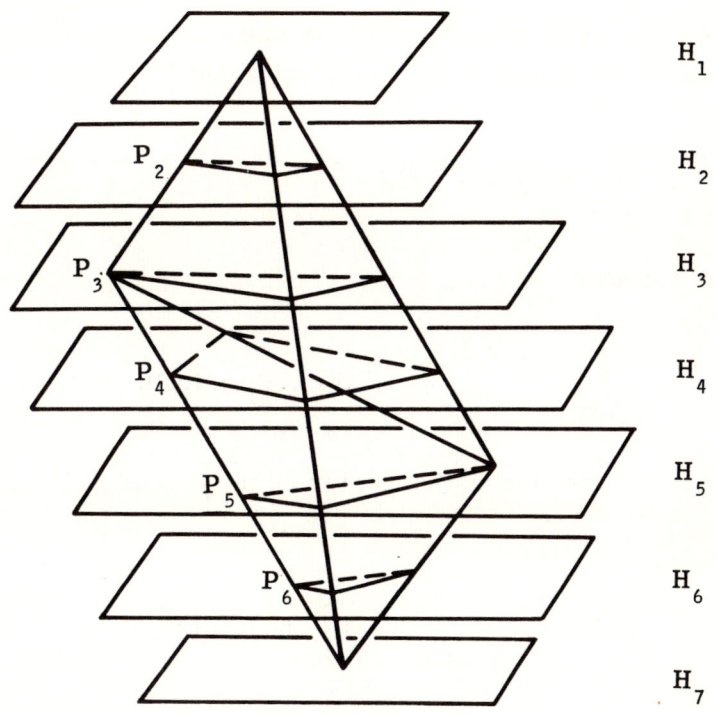

Figure 16

For each j-face F^j of P $(1 \leq j \leq d - 1)$, and each P_i $(2 \leq i \leq 2v - 2)$, define

$$\psi(F^j, P_i) = \begin{cases} 0 & \text{if } P_i \cap \text{relint } F^j = \emptyset, \\ 1 & \text{if } P_i \cap \text{relint } F^j \neq \emptyset. \end{cases}$$

For each F^j ($1 \le j \le d - 1$), the first hyperplane H_i to meet F^j will have odd suffix $i = 2l - 1$ (say), since it contains a single vertex of F^j, and therefore also of P. Similarly, the last hyperplane H_i to meet F^j will have odd suffix $i = 2m - 1$, and further, since $j \ge 1$, $l \ne m$. Consequently, for $i = 2l, \ldots, 2m - 2$, $P_i \cap \text{relint } F^j \ne \emptyset$, and $P_i \cap F^j$ is a $(j-1)$-face of P_i (compare Theorem 2 of Chapter 1). Hence for each such j-face of P, $\psi(F^j, P_i)$ takes the value 1 for one more even value of i than it does for odd values of i. This implies that

$$\sum_{i=2}^{2v-2} (-1)^i \psi(F^j, P_i) = 1,$$

and therefore

$$\sum_{j\text{-faces}} \sum_{i=2}^{2v-2} (-1)^i \psi(F^j, P_i) = f_j(P), \quad (1 \le j \le d-1)$$

where the first summation is over all the j-faces of P. Consequently,

$$\sum_{j=1}^{d-1} (-1)^j \sum_{j\text{-faces}} \sum_{i=2}^{2v-2} (-1)^i \psi(F^j, P_i) = \sum_{j=1}^{d-1} (-1)^j f_j(P).$$

(1)

The essence of the proof is to find an alternative expression for the term on the left in (1), by changing the order of summation. Noticing that if i is even or $j > 1$, then each $(j-1)$-face of P_i is the intersection of a j-face of P with H_i, whereas for i odd and $j = 1$, one vertex of P_i is a vertex of P, and each remaining vertex is the intersection of an edge of P with H_i, we deduce

$$\sum_{j\text{-faces}} \psi(F^j, P_i) = \begin{cases} f_0(P_i) - 1, & \text{if } j = 1 \text{ and } i \text{ odd}, \\ f_{j-1}(P_i), & \text{otherwise}. \end{cases}$$

Hence

$$\sum_{j=1}^{d-1} (-1)^j \sum_{j\text{-faces}} \psi(F^j, P_i) = \begin{cases} \sum_{j=1}^{d-1} (-1)^j f_{j-1}(P_i) + 1, & \text{if } i \text{ odd} \\ \sum_{j=1}^{d-1} (-1)^j f_{j-1}(P_i), & \text{if } i \text{ even} \end{cases}$$

$$= \begin{cases} (-1)^{d-1}, & \text{if } i \text{ odd}, \\ (-1)^{d-1} - 1, & \text{if } i \text{ even}. \end{cases}$$

The last equality holds because of our inductive hypothesis that Euler's Theorem holds for each of the $(d-1)$-polytopes P_i ($i = 2, \ldots, 2v-2$). It follows that

$$\sum_{i=2}^{2v-2} (-1)^i \sum_{j=1}^{d-1} (-1)^j \sum_{j\text{-faces}} \psi(F^j, P_i) =$$

$$(-1)^{d-1} - 1 - (v-2).$$

Using (1), and remembering that $v = f_0(P)$, we obtain

$$\sum_{j=1}^{d-1} (-1)^j f_j(P) = 1 + (-1)^{d-1} - f_0(P),$$

which is the Euler equation for P. This completes the proof of Theorem 18.

Euler's Theorem is the strongest possible assertion, in the sense that the Euler equation is the only linear relation between the numbers $f_j(P)$ that holds for all d-polytopes P. Geometrically, this means that if we define the f-vector $f(P)$ of P by

$$f(P) = (f_0(P), \ldots, f_{d-1}(P)) \in E^d ,$$

then the f-vectors of all d-polytopes P lie on the Euler hyperplane in E^d, defined by the Euler equation, and on no affine subspace of smaller dimension.

We shall prove this by induction on the dimension d, for the assertion is trivially true when $d = 1$ ($f(P) = (2)$) (and, in fact, for $d = 2$ also, where $f(P) = (n, n)$, $n \geq 3$). So, we make the hypothesis that the result holds in $d - 1$ dimensions ($d \geq 2$). Let

$$\sum_{j=0}^{d-1} \lambda_j f_j(P) = \mu \tag{2}$$

be any linear relation that holds for all d-polytopes P. We shall show that it must be a multiple of the Euler equation for P.

Let Q be any $(d - 1)$-polytope, and let R be the d-pyramid with basis Q, and S the d-bipyramid with basis Q. Then by Proposition 4 of §2.3 we have

$$f(R) = (1 + f_0(Q), f_0(Q) + f_1(Q), \ldots, f_{d-2}(Q) + 1) ,$$

and by Proposition 8,

$$f(S) = (2 + f_0(Q), 2f_0(Q) + f_1(Q), \ldots, 2f_{d-2}(Q)) .$$

Substituting in (2), and subtracting the first equation from the second, we obtain

$$\sum_{j=0}^{d-2} \lambda_{j+1} f_j(Q) = \lambda_{d-1} - \lambda_0 ,$$

which, by the inductive hypothesis, must be a multiple of

$$\sum_{j=0}^{d-2} (-1)^j f_j(Q) = 1 + (-1)^{d-2} .$$

Except in the trivial case when (2) is identically zero, we deduce that $\lambda_j = (-1)^{j-1} \lambda_1$ for $1 \le j \le d-1$, from which it follows that $\lambda_0 = (-1)^{d+1} \lambda_{d-1}$. Substituting the values of $f_j(P)$ for any polytope P in (2), we obtain $\mu = (1 + (-1)^{d-1})\lambda_0$. Hence (2) is a scalar multiple of the Euler equation (1), which proves our assertion.

On the other hand, the f-vectors of certain special classes of d-polytopes satisfy linear relations independent of the Euler equation. Here we shall investigate one case only, that of simplicial polytopes (§2.3 (v)).

Theorem 19. <u>Let P be a simplicial d-polytope. Then the numbers</u> $f_j(P)$ <u>of its j-faces</u> $(0 \le d \le -1)$ <u>satisfy, in addition to the Euler equation, the linear relations</u>

$$\sum_{j=k}^{d-1} (-1)^j \binom{j+1}{k+1} f_j(P) = (-1)^{d-1} f_k(P) \qquad (E_k^d)$$

<u>for</u> $k = 0, \ldots, d-2.$

With the usual convention $f_{-1}(P) = 1$, then the Euler equation formally corresponds to the case $k = -1$.

Proof. For each k-face F^k and j-face F^j of P ($0 \leq k \leq j \leq d - 1$), we define the function $\phi(F^k, F^j)$ by

$$\phi(F^k, F^j) = \begin{cases} 0 & \text{if } F^k \not\subseteq F^j \\ 1 & \text{if } F^k \subseteq F^j . \end{cases}$$

We prove the theorem by evaluating in two ways the sum

$$\sum_{j=k}^{d-1} (-1)^j \sum_{\text{k-faces}} \sum_{\text{j-faces}} \phi(F^k, F^j) , \qquad (3)$$

where the second and third sums are over all the k-faces and j-faces of P.

Firstly, from the proof of Theorem 16 (§2.2), we see that each face F^j of P such that $\phi(F^k, F^j) = 1$ corresponds to a $(j - k - 1)$-face of the $(d - k - 1)$-polytope P/F^k, and conversely. Hence $\sum_{\text{j-faces}} \phi(F^k, F^j)$ is precisely the number of $(j - k - 1)$-faces of P/F^k, and so

$$\sum_{j=k}^{d-1} (-1)^j \sum_{\text{j-faces}} \phi(F^k, F^j) = (-1)^{d-1}$$

by Theorem 18 applied to P/F^k (note that the value $j = k$ corresponds to the improper face $\emptyset = F^k/F^k$). Hence

$$\sum_{\text{k-faces}} \sum_{j=k}^{d-1} (-1)^j \sum_{\text{j-faces}} \phi(F^k, F^j) = (-1)^{d-1} f_k(P) . \qquad (4)$$

On the other hand, $\sum_{\text{k-faces}} \phi(F^k, F^j)$ is just the number of k-faces of the j-polytope F^j, and since each F^j is a j-simplex ($j \le d - 1$), we have

$$\sum_{\text{k-faces}} \phi(F^k, F^j) = \binom{j+1}{k+1}.$$

Hence

$$\sum_{\text{j-faces}} \sum_{\text{k-faces}} \phi(F^k, F^j) = \binom{j+1}{k+1} f_j(P);$$

substituting this in (3), and using (4), we immediately obtain

$$\sum_{j=k}^{d-1} (-1)^j \binom{j+1}{k+1} f_j(P) = (-1)^{d-1} f_k(P),$$

which for $k = 0, \ldots, d - 2$ are the equations (E_k^d) of the theorem.

The equations (E_k^d) ($k = -1, \ldots, d - 2$), the first of which is the Euler equation, are known as the <u>Dehn-Sommerville equations</u>. They can be conveniently written as the single algebraic identity

$$f(P, t) = (-1)^d f(P, 1 - t), \qquad (5)$$

where we write $f(P, t) = 1 - t f_0(P) + t^2 f_1(P) - \ldots + (-t)^d f_{d-1}(P)$. If we equate coefficients of t^{k+1} on each side of (5), we recover the equation (E_k^d).

Let us consider, for example, the cases $d = 3$ and $d = 4$. For $d = 3$ the Dehn-Sommerville equations are

$$-f_{-1}(P) + f_0(P) - f_1(P) + f_2(P) = f_{-1}(P), \qquad (E_{-1}^3)$$

$$f_0(P) - 2f_1(P) + 3f_2(P) = f_0(P) , \qquad (E_0^3)$$

$$-f_1(P) + 3f_2(P) = f_1(P) , \qquad (E_1^3)$$

of which the second and third are identical. For $d = 4$, the equations are

$$-f_{-1}(P) + f_0(P) - f_1(P) + f_2(P) - f_3(P) = -f_{-1}(P) , \qquad (E_{-1}^4)$$

$$f_0(P) - 2f_1(P) + 3f_2(P) - 4f_3(P) = -f_0(P) , \qquad (E_0^4)$$

$$-f_1(P) + 3f_2(P) - 6f_3(P) = -f_1(P) , \qquad (E_1^4)$$

$$f_2(P) - 4f_3(P) = -f_2(P) , \qquad (E_2^4)$$

of which the third and fourth are identical, and the first three are linearly dependent. The examples illustrate the following result:

Theorem 20. <u>For any</u> $d \geq 1$, <u>exactly</u> $[\tfrac{1}{2}(d+1)]$ <u>of the Dehn-Sommerville equations</u> (E_k^d) ($k = -1, \ldots, d-2$) <u>are independent. Geometrically, this means that the f-vectors</u> $f(P)$ <u>of simplicial polytopes</u> P <u>lie on an affine subspace of</u> E^d <u>of dimension</u> $[\tfrac{1}{2}d]$, <u>and on no affine subspace of lower dimension.</u>

Proof. It is easy to see that at least $[\tfrac{1}{2}(d+1)]$ of the Dehn-Sommerville equations are independent. For, if d is even, then for $j = 0, \ldots, \tfrac{1}{2}d - 1$, the term $f_{2j}(P)$ occurs in only the first $j + 1$ of the equations (E_{-1}^d), (E_1^d), $\ldots$, (E_{d-3}^d), and so they are independent. If d is odd, then (E_{-1}^d) is the only non-homogeneous equation, and for $j = 1, \ldots, \tfrac{1}{2}(d-1)$, the term f_{2j-1} occurs only in the first $j + 1$ of the equations (E_{-1}^d), (E_1^d), $\ldots$, (E_{d-2}^d), which are therefore independent.

To show that, at most, $[\frac{1}{2}(d+1)]$ of the equations are independent, it is enough to find $[\frac{1}{2}d]+1$ simplicial d-polytopes whose f-vectors are affinely independent. For this purpose we select the cyclic polytopes $C(v, d)$, $C(v+1, d)$, ..., $C(v+n, d)$ ($n = [\frac{1}{2}d]$), where v is any integer satisfying $v \geq d+1$. (See §2.3 (vi).) The $(n+1) \times d$ matrix whose i-th row is the vector $(1, f(C(v+i-1, d)))$ has leading minor

$$D(v, d) = \begin{pmatrix} 1 & \binom{v}{1} & \binom{v}{2} & \cdots & \binom{v}{n} \\ 1 & \binom{v+1}{1} & \binom{v+1}{2} & \cdots & \binom{v+1}{n} \\ \vdots & \cdots & \cdots & \cdots & \cdots \\ 1 & \binom{v+n}{1} & \binom{v+n}{2} & \cdots & \binom{v+n}{n} \end{pmatrix}$$

(see Proposition 19 of §2.3, or, recalling that $C(v+i-1, d)$ is n-neighbourly, Proposition 22). It is easily verified that $\det D(v, d) = 1$ (by continually subtracting each row from the row below), and so the vectors $f(C(v, d))$, ..., $f(C(v+n, d))$ are affinely independent. This completes the proof of Theorem 20.

Since about half of the Dehn-Sommerville equations are independent, we can solve the equations for about half the numbers $f_j(P)$ in terms of the other half. More precisely, in Chapter 4 we shall require expressions for $f_n(P)$, ..., $f_{d-1}(P)$ (we shall use the notation $n = [\frac{1}{2}d]$ throughout). There are several ways in which these expressions can be found; here we shall adapt a process devised by I. G. Macdonald [4], which yields the following.

Theorem 21. Let P be a simplicial d-polytope, and for $j = -1, \ldots, d - 1$ write $f_j = f_j(P)$ for the number of its j-faces (with $f_{-1} = 1$). Then

(i) If $d = 2n$ is even, for $0 \le p \le n - 1$,

$$f_{n+p} = \sum_{q=0}^{n-1} (-1)^q \frac{n-q}{p+q+1} \chi(n-1, p, q) f_{n-q-1} ; \quad (6)$$

(ii) If $d = 2n + 1$ is odd, for $0 \le p \le n$,

$$f_{n+p} = \sum_{q=0}^{n} (-1)^q \frac{n+p+2}{p+q+1} \chi(n, p, q) f_{n-q-1} ; \quad (7)$$

where

$$\chi(n, p, q) = \sum_{s \ge 0} \binom{n-s}{p} \binom{n-s+q+1}{n+1} , \quad (8)$$

and we adopt the usual convention that the binomial coefficient $\binom{a}{b}$ is zero if $b < 0$ or $b > a$.

Proof. Consider first the case $d = 2n$. If we define

$$f(t) = 1 - f_0 t + f_1 t^2 - \ldots + f_{2n-1} t^{2n} , \quad (9)$$

then the Dehn-Sommerville equations imply $f(t) = f(1 - t)$, as in (5). Equating the coefficients of t^{2n-1} on each side of this equation, we obtain $f_{2n-2} = n f_{2n-1}$. Consequently

$$g(t) = f(t) - (-1)^n f_{2n-1} t^n (1 - t)^n$$

is a polynomial of degree at most $2n - 2$, which clearly satisfies the same functional equation $g(t) = g(1 - t)$. Repeating this process at most n times leads us to a constant polynomial, and so we deduce that $f(t)$ can be written in the form $f(t) = F(u)$, where

$$F(u) = 1 - b_0 u + b_1 u^2 - \ldots + (-1)^n b_{n-1} u^n, \qquad (10)$$

and $u = t(1 - t)$. (The constant term comes from substituting the value $t = 0$.) Comparing (9) and (10), and equating the coefficients on each side, leads us to

$$f_r = b_r + \binom{r}{1} b_{r-1} + \binom{r-1}{2} b_{r-2} + \ldots, \qquad (11)$$

with the convention that $b_{-1} = 1$, $b_i = 0$ for $i < -1$ or $i > n - 1$, and the usual conventions for binomial coefficients. In particular, in the case $r = n + p$ ($p \geq 0$), this can be written

$$f_{n+p} = \sum_{k \geq 0} \binom{n-k}{p+k+1} b_{n-k-1}. \qquad (12)$$

We now solve the equations (11) (for $r = 0, \ldots, n - 1$) to express $b_0, \ldots, b_{n-1}$ in terms of $f_0, \ldots, f_{n-1}$. To do this we note that from the expression (10), $(-1)^{r+1} b_r$ is the residue at $u = 0$ of $F(u)/u^{r+2}$, which is

$$(-1)^{r+1} b_r = \frac{1}{2\pi i} \int_C \frac{F(u)}{u^{r+2}} du,$$

where C is some small circuit around the origin $u = 0$. Changing the variable in the integral by writing $u = t(1 - t)$, so that $du = (1 - 2t)dt$, we obtain for $r = 0, \ldots, n - 1$,

$$(-1)^{r+1} b_r = \frac{1}{2\pi i} \int_{C'} \frac{f(t)(1-2t)}{t^{r+2}(1-t)^{r+2}} \, dt,$$

for some other small circuit C' around the origin, which is the residue at $t = 0$ of $f(t)(1-2t)/t^{r+2}(1-t)^{r+2}$. In other words, using the notation $A(t)\,]t^r$ for the coefficient of t^r in $A(t)$,

$$(-1)^{r+1} b_r = \frac{f(t)(1-2t)}{(1-t)^{r+2}} \,] t^{r+1}$$

$$= \frac{f(t)}{(1-t)^{r+1}} \,] t^{r+1} - \frac{f(t)}{(1-t)^{r+2}} \,] t^r,$$

and so

$$b_r = \sum_{i=0}^{r} (-1)^i \binom{r+i}{i} f_{r-i} + \sum_{i=0}^{r-1} (-1)^i \binom{r+i+1}{i} f_{r-i-1}$$

$$= \sum_{i=0}^{r} (-1)^i \frac{r-i+1}{r+1} \binom{r+i}{i} f_{r-i}. \qquad (13)$$

Substituting (13) in (12) yields

$$f_{n+p} = \sum_{k \geq 0} \binom{n-k}{p+k+1} \sum_{i=0}^{n-k-1} (-1)^i \frac{n-k-i}{n-k} \binom{n-k+i-1}{i} f_{n-k-i-}$$

$$= \sum_{q=0}^{n-1} (-1)^q \{ \sum_{k \geq 0} (-1)^k \binom{n-k}{p+k+1} \frac{n-q}{n-k} \binom{n+q-1-2k}{q-k} \} f_{n-q-1}$$

$$= \sum_{q=0}^{n-1} (-1)^q \frac{n-q}{p+q+1} \{ \sum_{k \geq 0} (-1)^k \binom{p+q+1}{p+k+1} \binom{n+q-1-2k}{p+q} \} f_{n-q-}$$

$$(14)$$

In the case $d = 2n + 1$, the polynomial

$$f(t) = 1 - f_0 t + f_1 t^2 - \ldots - f_{2n} t^{2n+1}$$

satisfies $f(t) + f(1 - t) = 0$ (by (5)). This implies that $f(t)$ has a zero at $t = \frac{1}{2}$, and so it can be written in the form $f(t) = (1 - 2t)h(t)$, where $h(t)$ is a polynomial of degree $2n$ satisfying $h(t) = h(1 - t)$. Hence $h(t) = F(u)$, where $F(u)$ is again of the form (10). Comparing the two expressions for $f(t)$, and equating coefficients of t^{r+1}, we have

$$f_r = b_r + \frac{r+2}{r+1} \binom{r+1}{1} b_{r-1} + \frac{r+2}{r} \binom{r}{2} b_{r-2} + \ldots, \quad (15)$$

where we adopt the same conventions for b_i as in (11). In particular, if $r = n + p$ ($p \geq 0$), we obtain the analogue of (12)

$$f_{n+p} = \sum_{k \geq 0} \frac{n+p+2}{n-p-2k} \binom{n-k}{p+k+1} b_{n-k-1}. \quad (16)$$

As before, we solve (15) (for $r = 0, \ldots, n - 1$) to express $b_0, \ldots, b_{n-1}$ in terms of $f_0, \ldots, f_{n-1}$. Here the computation is easier, and yields

$$(-1)^{r+1} b_r = \frac{f(t)}{(1-t)^{r+2}} \Big] t^{r+1},$$

which for $r = 0, \ldots, n - 1$ is

$$b_r = \sum_{i=0}^{r+1} (-1)^i \binom{r+i+1}{i} f_{r-i}, \quad (17)$$

with the usual convention $f_{-1} = 1$. Substituting (17) in (16) gives

$$f_{n+p} = \sum_{k \geq 0} \frac{n+p+2}{n-p-2k} \binom{n-k}{p+k+1} \sum_{i=0}^{n-k} (-1)^i \binom{n-k+i}{i} f_{n-i-k-1}$$

$$= \sum_{q=0} (-1)^q \{ \sum_{k \geq 0} (-1)^k \frac{n+p+2}{n-p-2k} \binom{n-k}{p+k+1} \binom{n+q-2k}{q-k} \} f_{n-q-1}$$

$$= \sum_{q=0}^{n} (-1)^q \frac{n+p+2}{p+q+1} \{ \sum_{k \geq 0} (-1)^k \binom{p+q+1}{p+k+1} \binom{n+q-2k}{p+q} \} f_{n-q-1}.$$

(18)

If we write

$$\chi(n, p, q) = \sum_{k \geq 0} (-1)^k \binom{p+q+1}{p+k+1} \binom{n+q-2k}{p+q} \quad (19)$$

we see that (14) and (18) are just (6) and (7) respectively. So, to complete the proof of Theorem 21, it just remains to show that the expressions (8) and (19) for $\chi(n, p, q)$ are equivalent. We take $\chi(n, p, q)$ in the form (19). Using the relations

$$\binom{a}{b} = \binom{a-1}{b} + \binom{a-1}{b-1}, \quad (20)$$

$$\binom{a}{b}\binom{a+c}{a} = \binom{b+c}{b}\binom{a+c}{b+c}, \quad (21)$$

(we have in fact already used (21) above), we may rewrite $\chi(n, p, q)$ as

$$X(n, p, q) = \sum_{k \geq 0} (-1)^k \{ \binom{p+q}{p+k+1} + \binom{p+q}{p+k} \} \binom{n+q-2k}{p+q}$$

$$= \sum_{k \geq 0} (-1)^k \{ \binom{n-k+1}{p+k+1} \binom{n+q-2k}{n-k+1} + \binom{n-k}{p+k} \binom{n+q-2k}{n-k} \}. \tag{22}$$

We now use the following relation, which follows directly from (20),

$$\binom{a}{b} = \sum_{v \geq 0} \binom{a-v-1}{b-1}, \tag{23}$$

and write (22) as

$$\sum_{k \geq 0} (-1)^k \{ \sum_{v \geq 0} \sum_{w \geq 1} \binom{n-k-v}{p+k} \binom{n+q-2k-w}{n-k} + \binom{n-k}{p+k} \binom{n+q-2k}{n-k} \}. \tag{24}$$

As a temporary notation, let us write

$$\psi(k, v, w) = (-1)^k \binom{n-k-v}{p+k} \binom{n+q-2k-w}{n-k}, \tag{25}$$

so that (24) is

$$X(n, p, q) = \sum_{k \geq 0} \{ \sum_{v \geq 0} \sum_{w \geq 1} \psi(k, v, w) + \psi(k, 0, 0) \},$$

which can clearly be rearranged as

$$\sum_{k \geq 0} \sum_{v \geq 0} \sum_{w \geq v} \psi(k, v, w) + \sum_{k \geq 0} \sum_{w \geq 1} \sum_{v \geq w+1} \psi(k, v, w). \tag{26}$$

We now express (26) alternatively as follows. In the first term we put $s = k + v$, and in the second $s = k + w$, and rearrange the sums. (All the sums are finite, so no question of convergence arises.) We then obtain

$$\chi(n, p, q) = \sum_{s \geq 0} \{ \sum_{k=0}^{s} \sum_{w \geq s-k} \psi(k, s-k, w) +$$

$$\sum_{k=0}^{s-1} \sum_{v \geq s-k+1} \psi(k, v, s-k) \} . \qquad (27)$$

Substituting for $\psi(k, v, w)$ from (25), and using (23) again

$$\sum_{w \geq s-k} \psi(k, s-k, w) = \sum_{w \geq s-k} (-1)^k \binom{n-s}{p+k} \binom{n+q-2k-w}{n-k}$$

$$= (-1)^k \binom{n-s}{p+k} \binom{n+q+1-s-k}{n-k+1} . \qquad (28)$$

Similarly,

$$\sum_{v \geq s-k+1} \psi(k, v, s-k) = \sum_{v \geq s-k+1} (-1)^k \binom{n-k-v}{p+k} \binom{n+q-s-k}{n-k}$$

$$= (-1)^k \binom{n-s}{p+k+1} \binom{n+q-s-k}{n-k} \qquad (29)$$

$$= - (-1)^t \binom{n-s}{p+t} \binom{n+q-1-s-t}{n-t+1} , \qquad (30)$$

where we have put $t = k + 1$ in (29) to obtain (30). Substituting (28) and (30) in (27), and changing t to k, we have

$$\chi(n, p, q) = \sum_{s \geq 0} \{ \sum_{k=0}^{s} (-1)^k \binom{n-s}{p+k} \binom{n+q+1-s-k}{n-k+1} -$$

$$\sum_{k=1}^{s} (-1)^k \binom{n-s}{p+k} \binom{n+q+1-s-k}{n-k+1} \} = \sum_{s \geq 0} \binom{n-s}{p} \binom{n-s+q+1}{n+1},$$

which is just (8). This completes the proof of Theorem 21.

Theorem 21 shows that the numbers of r-faces ($[\frac{1}{2}d] \leq r \leq d-1$) of a simplicial $[\frac{1}{2}d]$-neighbourly d-polytope P are completely determined by the numbers of j-faces ($0 \leq j \leq [\frac{1}{2}d] - 1$), and since these numbers are

$$f_j(v, d) = \binom{v}{j+1}, \quad v = f_0(P), \quad 0 \leq j < [\tfrac{1}{2}d],$$

this shows that every such polytope P has the same number of faces of each dimension as the cyclic polytope C(v, d) with the same number of vertices. This provides an alternative method for calculating the numbers $f_j(v, d)$ of these faces (Proposition 19).

For certain special values of p the formulae (6) and (7) simplify as follows: For $d = 2n$,

$$f_n = \sum_{i=0}^{n-1} (-1)^{n-i-1} \frac{i+1}{n+1} \binom{2n-i}{n} f_i,$$

$$f_{2n-1} = \sum_{i=0}^{n-1} (-1)^{n-i-1} \frac{i+1}{n} \binom{2n-2-i}{n-1} f_i.$$

For $d = 2n + 1$,

$$f_n = \sum_{i=-1}^{n-1} (-1)^{n-i-1} \binom{2n-i+1}{n+1} f_i ,$$

$$f_{2n} = 2 \sum_{i=-1}^{n-1} (-1)^{n-i-1} \binom{2n-i-1}{n} f_i ,$$

with the usual convention $f_{-1} = 1$.

In Chapter 4 we shall also require the equations (6) and (7) explicitly in the cases $d = 3$, 4 and 5, and we conclude the section by reproducing these:

$$d = 3 \quad \begin{aligned} f_1 &= 3f_0 - 6 , \\ f_2 &= 2f_0 - 4 . \end{aligned}$$

$$d = 4 \quad \begin{aligned} f_2 &= 2f_1 - 2f_0 , \\ f_3 &= f_1 - f_0 . \end{aligned}$$

$$d = 5 \quad \begin{aligned} f_2 &= 4f_1 - 10f_0 + 20 , \\ f_3 &= 5f_1 - 15f_0 + 30 , \\ f_4 &= 2f_1 - 6f_0 + 12 . \end{aligned}$$

2.5 PULLING THE VERTICES OF A POLYTOPE

In the last section of this chapter we shall describe a method of changing the combinatorial type of a polytope P, known as 'pulling its vertices'. This process, which is of particular importance in connection with the Upper Bound Conjecture, has the effect of transforming P into a simplicial polytope with the same number of vertices as P, and at least as many faces of higher dimensions.

To begin with, we consider the effect of adding a new vertex to a polytope (a process which may result in other vertices being removed). Let P be a d-polytope in E^d, let H be a hyperplane in E^d such that H ∩ int P = ∅, and let w be any point of E^d. If w belongs to the open half-space bounded by H which contains P, then we say that w is <u>beneath</u> H (with respect to P); and if w is in the other open half-space, then w is said to be <u>beyond</u> H. Extending this terminology, we say that w is beneath or beyond a facet F of P when it is beneath or beyond the corresponding hyperplane aff F, respectively.

The following theorem describes the faces of the new polytope formed by adding a new vertex to a given polytope.

Theorem 22. <u>Let P be a d-polytope in E^d, and let $w \in E^d \setminus P$ be a given point. Then</u>

$$P' = conv(\{w\} \cup P)$$

<u>is also a d-polytope, and:</u>

(1) <u>A face F of P is a face of P' if and only if there is a facet F" of P which contains F, such that w is beneath F" with respect to P.</u>

(2) <u>If F is a face of P, then $F' = conv(\{w\} \cup F)$ is a face of P' if and only if either (a) w ∈ aff F, or (b) w is beyond at least one facet of P containing F and beneath another.</u>

<u>Moreover, each face of P' is of one and only one of the two types described above.</u>

Proof. Clearly w is a vertex of P'. Let F' be a face of P', and let H' be a supporting hyperplane of P' such that H' ∩ P' = F'. Then by Theorem 5, H' ∩ P = F is a face of P, and either H' does not contain w, in which case F' = F is a face of P, or H' contains w, and in this case F' = conv({w} ∪ F). We discuss the two cases separately.

A facet F" of P will be a facet of P' if and only if aff F" supports P' and w ∉ aff F"; that is, if and only if w is beneath F" with respect to P, and therefore also beneath F" with respect to P'. So, if F is a face of P contained in the facet F" of P, and if w is beneath F", then F is a face of P' (Theorem 4). Conversely, if F is a face of P' such that w ∉ F, then F is a face of P, and F is the intersection of all the facets of P which contain it (Theorem 9). Clearly w cannot be beyond all these facets (for this would imply that relint F ⊂ int P') and it cannot lie in all the hyperplanes determined by these facets (for this would imply that w ∈ aff F). We conclude that w is beneath at least one facet of P which contains F, which completes the proof of (1).

Now suppose that F is a face of P, and that F' = conv({w} ∪ F) is a face of P'. Then clearly F = P ∩ aff F'. Let x_0 ∈ relint F, y_0 ∈ int P, and E = aff {w, x_0, y_0}. We consider the intersection P_0 = E ∩ P of the (2-dimensional) plane E with P. By Corollary 2 to Theorem 4, P_0 is a convex polygon (see Figures 17 and 18). The line L = aff {w, x_0} is the intersection of E with aff F', and so F_0 = L ∩ P_0 is either an edge of P_0 (as in Figure 17) or a vertex of P_0 (Figure 18). In the former case, w ∈ aff F_0 ⊆ aff F, and so we have case (a). In the latter case, w is (in E) beneath one and beyond the other of the edges of P_0 containing F_0 = {x_0}. If F_1 and F_2 are any two facets of P containing these two edges of P_0, it follows that (in E^d) w is beyond one of F_1 and F_2, and beneath the other, which is case (b).

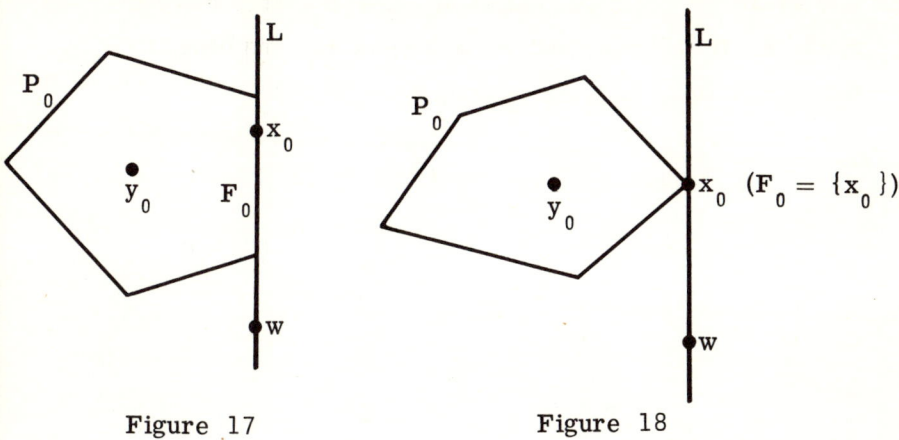

Figure 17   Figure 18

It remains to be shown that if F is a face of P satisfying either (a) or (b), then $F' = \operatorname{conv}(\{w\} \cup F)$ is a face of P'. The case (a) is trivial, since if H is a supporting hyperplane of P such that $H \cap P = F$, then because $w \in \operatorname{aff} F$ it follows that $H \cap P' = F'$. For the case (b), let F be a face of P, and let F_1 and F_2 be two facets of P such that $F \subseteq F_1 \cap F_2$, and w is beyond F_1 and beneath F_2. Suppose that $o \in F$, let H_0 be a supporting hyperplane of P such that $H_0 \cap P = F$, and for $i = 1, 2$, let $H_i = \operatorname{aff} F_i$, say

$$H_i = \{x \in E^d \mid \langle x, a_i \rangle = 0\},$$

where (without loss of generality),

$$P \subset H_i^+ = \{x \in E^d \mid \langle x, a_i \rangle \geq 0\} \quad (i = 0, 1, 2).$$

If we rotate the hyperplanes H_i slightly about $H_i \cap H_0$, towards H_0, we obtain new hyperplanes H_1' and H_2', such that for $i = 1, 2$, $H_i' \cap P = F$, and w is beyond H_1' and beneath H_2'. Explicitly, let

$$\eta_1 = \sup\{\eta \mid \langle w, a_1 + \eta a_0 \rangle \leq 0\} > 0,$$

$$\eta_2 = \sup\{\eta \mid \langle w, a_2 + \eta a_0 \rangle \geq 0\} > 0,$$

and let

$$H_i' = \{x \in E^d \mid \langle x, a_i + \tfrac{1}{2}\eta_i a_0 \rangle = 0\}, \quad i = 1, 2.$$

This implies that the hyperplane $H_0' = \mathrm{aff}(\{w\} \cup (H_1' \cap H_2'))$ contains w, and satisfies $H_0' \cap P = F$. Thus

$$H_0' \cap P' = H_0' \cap \mathrm{conv}(\{w\} \cup P) = \mathrm{conv}(\{w\} \cup F) = F'$$

is a face of P', as claimed. This completes the proof of Theorem 22.

We are now in a position to define the process which we call 'pulling'. Let P be a d-polytope in E^d, and let w be a vertex of P. Let $w' \in E^d \setminus P$ be such that the half-open segment $\mathrm{conv}\{w, w'\} \setminus \{w\}$ does not meet any hyperplane spanned by vertices of P. If w is an interior point of $P' = \mathrm{conv}(\{w'\} \cup P)$, then we say that P' is obtained from P by <u>pulling</u> w to w'. The conditions on w' just stated ensure that w' is beyond all the facets of P which contain w, and beneath all the other facets of P.

The following result is a consequence of Theorem 22.

Theorem 23. Let P' be a d-polytope in E^d obtained from the d-polytope P by pulling the vertex w of P to w'. Then for $k = 1, \ldots, d - 1$, the k-faces of P' are as follows:

(1) The k-faces of P which do not contain w.

(2) The convex hulls of the form $F' = \operatorname{conv}(\{w'\} \cup F)$, where F is a $(k-1)$-face not containing w of a facet of P which does contain w.

The second part of Theorem 23 implies immediately:

Theorem 24. In the notation of Theorem 23, for $k = 1, \ldots, d - 1$, each k-face of P' which contains w' is a k-pyramid with apex w'. Therefore $f_0(P') = f_0(P)$, and for $k = 1, \ldots, d - 1$, $f_k(P') \geq f_k(P)$.

Bearing in mind Theorem 24, we also deduce from Theorem 23:

Theorem 25. If the polytope Q is obtained from P by successively pulling each of the vertices of P, then Q is a simplicial d-polytope satisfying $f_0(Q) = f_0(P)$, $f_k(Q) \geq f_k(P)$ ($k = 1, \ldots, d - 1$). Moreover, if some j-face of P is not a simplex, then for each $k = j - 1, \ldots, d - 1$, $f_k(Q) > f_k(P)$ (with strict inequality).

These results can be summarized as follows: If we pull all the vertices of P to make it into a simplicial polytope, then we do not alter the number of its vertices, but tend to increase the number of its j-faces for $j = 1, \ldots, d - 1$. Further, one of these numbers will actually increase unless P was simplicial to start with. Hence if we are interested in finding d-polytopes with a given number of vertices, and a maximum number of faces of other dimensions (as we shall be in our discussion of the Upper Bound Conjecture), then we shall lose no generality in restricting our attention to simplicial polytopes.

In [12] a stronger form of Theorem 25 is proved, in which lower bounds for $f_k(Q) - f_k(P)$ are established when P is not simplicial.

3. Gale Diagrams, and Polytopes with Few Vertices

In proving some of the special cases of the Upper Bound Conjecture, it is necessary to investigate the combinatorial structure of d-polytopes with 'few' vertices, that is, with $d + 2$ or $d + 3$ vertices. In order to do this, we shall make use of the technique of representing a polytope by means of a Gale diagram. This idea, which we introduce for its intrinsic interest as much as for its applications, originated with David Gale (1956), although (with the benefit of hindsight) many foretastes of the concept appeared previously. Micha Perles (1966) was the first to realize the full power of the method.

Briefly, if P is a d-polytope in E^d with n vertices, then a Gale diagram of P consists of a set of n points in E^{n-d-1}, in one-to-one correspondence with the vertices of P. From the Gale diagram it is possible to determine all the combinatorial properties of P; that is, those subsets of the vertices of P which define faces of P, the combinatorial types of these faces, and so on. Of particular importance is the fact that, if n is not much larger than d (in fact, if $n \leq 2d$), then the dimension of the Gale diagram is smaller than that of P.

There are two approaches to Gale diagrams, one geometric and the other algebraic. In either case we define first what is known as a Gale transform, from which a Gale diagram is afterwards derived. Here we shall describe both methods; our treatment follows closely that of a recent paper [6] by the authors.

3.1 GALE TRANSFORMS: GEOMETRIC FORMULATION

Let P be a given d-polytope in E^d with n vertices $x_1, \ldots, x_n$, and suppose (without loss of generality) that $o \in \operatorname{int} P$. Let P^* be the polar set of P, and let T^{n-1} be a regular $(n-1)$-simplex in E^{n-1}, the centroid of whose vertices is the origin $o \in E^{n-1}$. By Theorem 12 of Chapter 2, there is a d-dimensional affine subspace L^d of E^{n-1} through o, such that the section $L^d \cap T^{n-1}$ is projectively equivalent to P^*, and such that the origins of E^d and E^{n-1} correspond under the projective equivalence. Let L^{n-d-1} be the $(n-d-1)$-dimensional affine subspace of E^{n-1} that passes through o and is perpendicular to L^d, and let Π^{n-d-1} denote orthogonal projection onto L^{n-d-1}. Then we call $\overline{V} = (\operatorname{vert} T^{n-1}) \Pi^{n-d-1} \subset L^{n-d-1}$, the image of the set of vertices of T^{n-1} under this orthogonal projection, a Gale transform of the set $\operatorname{vert} P$ of the vertices of P.

The polar set of T^{n-1} in E^{n-1} is clearly also a regular $(n-1)$-simplex with the same centroid o; in fact $(T^{n-1})^* = -\lambda T^{n-1}$, for some $\lambda > 0$. Since we have made the origins of E^d and E^{n-1} correspond, the projective equivalence used above is of the form

$$x\Theta = \frac{x\Lambda}{1 - \langle x, p \rangle},$$

where Λ is a linear map: $E^d \to E^{n-1}$, and $p \in \operatorname{int} P$. Using Π^d to denote orthogonal projection onto L^d, Theorems 14 and 15 of Chapter 2 imply that $(T^{n-1})^* \Pi^d = P'$ is affinely (and not merely projectively) equivalent to P. We have proved:

Lemma 1. <u>Any polytope with n vertices is affinely equivalent to the image of a regular (n - 1)-simplex under orthogonal projection.</u>

Thus (vert T^{n-1})*Π^{n-d-1} = $-\lambda \overline{V}$ is linearly equivalent to the Gale transform $\overline{V}$. Let the vertices of T^{n-1} be $y_1, \ldots, y_n$, labelled in such a way that, for $i = 1, \ldots, n$

$$(-\lambda y_i)\Pi^d = x_i'$$

is the vertex of P' corresponding to the vertex x_i of P. We then denote

$$y_i \Pi^{n-d-1} = \overline{x}_i \quad (i = 1, \ldots, n),$$

so that $\overline{V} = \{\overline{x}_1, \ldots, \overline{x}_n\}$.

We notice that the points of $\overline{V}$ may not all be distinct; if not we must regard the Gale transform as including points of multiplicity greater than one. We shall see shortly that the position of the origin o in L^{n-d-1}, relative to the points of $\overline{V}$, is of vital importance.

It is clear that the construction of a Gale transform as described above is, to some extent, arbitrary. For example, we can choose any polar set of P to exhibit as a section of T^{n-1}, and each such choice will, in general, lead to a different Gale transform.

Before stating the main result of this section, we introduce some useful terminology and notation: A subset $Z \subseteq V = $ vert P is called a <u>coface</u> of P if F = conv (V\Z) is a face of P. For any subset $Z \subseteq V$, we write $\overline{Z} \subseteq \overline{V}$ for the set of transforms $\overline{x}_i$ of the points x_i in Z.

Theorem 1. A subset $Z \subseteq \text{vert } P$ is a coface of the polytope P if and only if, in a Gale transform of vert P,

$$o \in \text{relint conv } \overline{Z} \ .$$

Proof. Without loss of generality identify P^* and $L^d \cap T^{n-1}$. To each vertex x_i of P corresponds a facet F_i of P^*, and from the definitions

$$F_i = L^d \cap D_i \ , \qquad (1)$$

where

$$D_i = \text{conv } \{y_1, \ldots, y_{i-1}, y_{i+1}, \ldots, y_n\}, \quad 1 \le i \le n \ ,$$

is a facet of T^{n-1}. Let $Z = \{x_1, \ldots, x_r\}$ be an arbitrary subset (after suitable renumbering) of vert P. If Z is a coface of P, then conv $\{x_{r+1}, \ldots, x_n\}$ is a face of P, and so

$$\hat{F} = F_{r+1} \cap \ldots \cap F_n$$

is a face of P^*. $F_{r+1}, \ldots, F_n$ are all the facets of P^* which contain $\hat{F}$. From (1) above it follows at once that

$$\hat{F} = L^d \cap D \ , \qquad (2)$$

where D is the face of T^{n-1},

$$D = D_{r+1} \cap \ldots \cap D_n = \text{conv } \{y_1, \ldots, y_r\} \ . \qquad (3)$$

Then D is the face of T^{n-1} of smallest dimension which gives rise to $\hat{F}$ in the manner of (2), so that

$$\text{relint } \hat{F} \subseteq \text{relint } D.$$

Remembering that $\hat{F} \subset L^d$, on applying Π^{n-d-1} we obtain

$$o \in (\text{relint } D)\Pi^{n-d-1} = \text{relint conv } \{y_1 \Pi^{n-d-1}, \ldots, y_r \Pi^{n-d-1}\}$$

$$= \text{relint conv } \overline{Z}.$$

Conversely, if $\overline{Z} \subseteq \overline{V}$ is such that $o \in \text{relint conv } \overline{Z}$, then defining D by (3), $L^d \cap D \neq \emptyset$ is a face of P^*, so that $F = \text{conv } \{x_{r+1}, \ldots, x_n\}$ is the corresponding face of P, and $Z = \{x_1, \ldots, x_r\}$ is a coface of P. This completes the proof of Theorem 1.

Before discussing the consequences of this theorem, so that the reader may have some idea of what Gale transforms are like, in Figure 19 we give two examples of polyhedra (3-polytopes) with 6 vertices, and the corresponding Gale transforms (which are, of course, in $n - d - 1 = 6 - 3 - 1 = 2$ dimensions). The reader should verify that Theorem 1 holds in these two cases; he may also find it useful to refer back to these examples as illustrations of future assertions about Gale transforms.

A Gale transform of the set of vertices of a d-simplex T^d has dimension $(d + 1) - d - 1 = 0$; it consists of $d + 1$ points coincident with the origin. This special case has to be excluded in our first corollary.

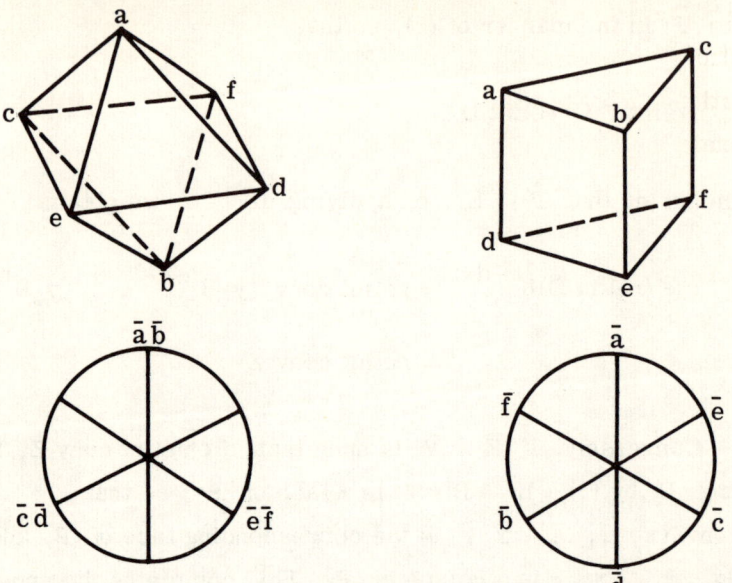

Figure 19

Corollary 1. <u>A Gale transform $\overline{V}$ of the set of vertices of a polytope (other than a simplex) has the property that every open half-space with o on its boundary contains at least two points of $\overline{V}$.</u>

Proof. Each vertex x_i of the polytope P is a face of P, so for $i = 1, \ldots, n$,

$$Z_i = \{x_1, \ldots, x_{i-1}, x_{i+1}, \ldots, x_n\}$$

is a coface of P, and by Theorem 1, this implies that

$$o \in \text{relint conv } \{\overline{x}_1, \ldots, \overline{x}_{i-1}, \overline{x}_{i+1}, \ldots, \overline{x}_n\}.$$

But this could not hold if some open half-space with o on its boundary contained x_i alone. Hence the corollary is true.

124

There is a partial converse to this corollary, stating the conditions under which a set of points is a Gale transform of the set of vertices of a polytope, but it is more convenient to prove this by algebraic means in §3.2.

The geometrical construction of a Gale transform leads immediately to a useful formula (4) for the dimension of a face F of P in terms of the transforms of the vertices belonging to the corresponding coface Z. Using the same notation as in the proof of Theorem 1, we consider the restriction of Π^{n-d-1} to the subspace aff D. This restriction is orthogonal projection onto a subspace orthogonal to $L^d \cap$ aff D, and so,

$$\dim D = \dim (L^d \cap D) + \dim (D\Pi^{n-d-1}) .$$

From this we obtain

$$\operatorname{card} \overline{Z} - 1 = \dim \hat{F} + \dim \overline{Z} ,$$

and so, since $\dim F + \dim \hat{F} = d - 1$

$$\dim F = d + \dim \overline{Z} - \operatorname{card} \overline{Z} . \qquad (4)$$

Two special cases of (4) are worthy of mention; they are stated as corollaries, since they are also consequences of Theorem 1.

Corollary 2. <u>If F is a facet of P and Z is the corresponding coface, then in a Gale transform of vert P, $\overline{Z}$ is the set of vertices of a (non-degenerate) simplex with o in its relative interior.</u>

Proof. In (4) we put dim $F = d - 1$, so that card $\overline{Z} =$ dim $\overline{Z} + 1$, which implies the statement of the corollary.

Alternatively, we note that if F is a face, and Z the corresponding coface, and if $\overline{Z}$ is not the set of vertices of a simplex, then card $\overline{Z} \geq$ dim $\overline{Z} + 2$. By Carathéodory's Theorem (Theorem 11 of Chapter 1) it follows that some proper subset of $\overline{Z}$ has the property that the origin lies in the relative interior of its convex hull. Hence some proper subset of Z is a coface of P, and therefore F is a proper face of some face of P; that is to say, F is not a facet of P.

Corollary 3. *If a face F of P is a simplex, and Z is the corresponding coface, then $\overline{Z}$ spans aff $\overline{V}$ positively.*

Proof. In (4) we put dim $F = s$, card $Z = n - s - 1$, and then dim $\overline{Z} = n - d - 1 =$ dim $\overline{V}$. Since Z is a coface, $o \in$ int conv Z, and the assertion follows.

Alternatively, we note that since F is a simplex, every subset of vert F is the set of vertices of a simplex which is a face of P. Hence every subset of $V =$ vert P which contains Z is a coface, and so every subset of $\overline{V}$ which contains $\overline{Z}$ has o in the relative interior of its convex hull. Clearly this condition cannot hold unless $\overline{Z}$ has the same dimension as $\overline{V}$, and the statement of the corollary follows.

The importance of Corollary 3 is that it enables us to recognize Gale transforms of the sets of vertices of simplicial polytopes. The condition may be restated thus:

Corollary 4. *A polytope P is simplicial if and only if, in a Gale transform $\overline{V}$ of vert P, for every hyperplane H containing o, we have*

$o \notin \text{relint conv}(\bar{V} \cap H)$.

3.2 GALE TRANSFORMS: ALGEBRAIC FORMULATION

The algebraic formulation of Gale transforms is, at first sight, completely different from that of the last section. In the next section we shall explain the connection between the two.

As before, we write

$$\text{vert } P = V = \{x_1, \ldots, x_n\}$$

for the set of vertices of the d-polytope P in E^d. In fact, the following treatment applies more generally to any set V of n points in E^d whose convex hull is d-dimensional; the special properties of the sets of vertices of polytopes will be explained later. We consider the set D(V) of affine dependences of V, that is, the set of all vectors $(\lambda_1, \ldots, \lambda_n) \in E^n$ such that

$$\left. \begin{array}{l} \lambda_1 x_1 + \ldots + \lambda_n x_n = o, \\ \lambda_1 + \ldots + \lambda_n = 0. \end{array} \right\} \quad (1)$$

It is trivial to check that D(V) is a vector space, and since the n points x_i span E^d affinely, the dimension of D(V) must be n - d - 1. We choose a basis $\{a_1, \ldots, a_{n-d-1}\}$ of D(V), and suppose that

$$a_j = (\alpha_{j1}, \ldots, \alpha_{jn}), \quad j = 1, \ldots, n-d-1.$$

Let

$$\underline{A}(V) = \begin{pmatrix} a_1 \\ \cdot \\ \cdot \\ \cdot \\ a_{n-d-1} \end{pmatrix} = \begin{pmatrix} \alpha_{11} & \cdots & \alpha_{1n} \\ \cdot & & \cdot \\ \cdot & & \cdot \\ \cdot & & \cdot \\ \alpha_{n-d-1,1} & \cdots & \alpha_{n-d-1,n} \end{pmatrix}$$

be the matrix whose rows are $a_1, \ldots, a_{n-d-1}$. For each $i = 1, \ldots, n$, let

$$\bar{x}_i = (\alpha_{1i}, \ldots, \alpha_{n-d-1,i}) \in E^{n-d-1}$$

be the i-th column of the matrix $\underline{A}(V)$. Then the set

$$\bar{V} = \{\bar{x}_1, \ldots, \bar{x}_n\} \subset E^{n-d-1}$$

is a Gale transform of V, with $\bar{x}_i$ corresponding to $x_i \in V$.

As in the geometric formulation, there is considerable arbitrariness in the construction of a Gale transform. For example, we may choose our basis of D(V) in many ways, and these lead to different (but linearly equivalent) Gale transforms. We begin by showing that the result corresponding to Theorem 1 holds for the Gale transform just described.

Theorem 1A. Let P be a polytope, and let V = vert P. Then the subset $Z \subseteq V$ is a coface of P if and only if, in a Gale transform $\bar{V}$ of V,

$$o \in \text{relint conv } \bar{Z} .$$

Proof. By Theorem 11 of Chapter 2, Z is a coface of P if and only if

$$\text{aff}(V \setminus Z) \cap \text{conv } Z = \emptyset. \qquad (2)$$

So, let us suppose that $Z = \{x_1, \ldots, x_r\}$ is not a coface of P. Then by (2),

$$\text{conv}\{x_1, \ldots, x_r\} \cap \text{aff}\{x_{r+1}, \ldots, x_n\} \neq \emptyset,$$

so that there is a point x, such that $x \in \text{conv}\{x_1, \ldots, x_r\}$, and $x \in \text{aff}\{x_{r+1}, \ldots, x_n\}$. From the first of these, we see that there exist scalars $\lambda_1, \ldots, \lambda_r$ such that

$$x = \sum_{i=1}^{r} \lambda_i x_i, \quad \sum_{i=1}^{r} \lambda_i = 1, \quad \lambda_i \geq 0 \ (i = 1, \ldots, r), \qquad (3)$$

so that at least one λ_i is strictly positive. From the second, there are scalars $\lambda_{r+1}, \ldots, \lambda_n$ such that

$$x = \sum_{i=r+1}^{n} (-\lambda_i) x_i, \quad \sum_{i=r+1}^{n} (-\lambda_i) = 1. \qquad (4)$$

From (3) and (4) we have

$$\sum_{i=1}^{n} \lambda_i x_i = 0, \quad \sum_{i=1}^{n} \lambda_i = 0, \qquad (5)$$

so that $l = (\lambda_1, \ldots, \lambda_n)$ is an affine dependence of the set V; that is, $l \in D(V)$. Consequently l can be written as a linear combination of the basis $\{a_1, \ldots, a_{n-d-1}\}$ of $D(V)$, say

$$\ell = \gamma_1 a_1 + \ldots + \gamma_{n-d-1} a_{n-d-1},$$

for some scalars $\gamma_1, \ldots, \gamma_{n-d-1}$. Put $c = (\gamma_1, \ldots, \gamma_{n-d-1}) \in E^{n-d-1}$. Then by the definition of the Gale transform $\overline{V}$,

$$\lambda_i = \langle c, \overline{x}_i \rangle, \quad i = 1, \ldots, n,$$

so that from (3)

$$\langle c, \overline{x}_i \rangle \geq 0, \quad i = 1, \ldots, r,$$

with strict inequality for at least one suffice i. Hence $\overline{x}_1, \ldots, \overline{x}_r$ lie in the closed half-space $\langle c, \overline{x} \rangle \geq 0$ of E^{n-d-1}, with at least one of the points in the corresponding open half-space, and so

$$o \notin \text{relint conv} \{\overline{x}_1, \ldots, \overline{x}_r\}. \tag{6}$$

It is easily seen that the argument is reversible, so that (6) is a necessary and sufficient condition for Z to be a coface of P. This completes the proof of Theorem 1A.

Clearly the assertions of Corollaries 1, 2, 3 and 4 of Theorem 1 also hold for Gale transforms defined algebraically. However, we can now prove a converse to Corollary 1, which answers the important question as to which sets of points in E^{n-d-1} are Gale transforms of the sets of vertices of polytopes.

Theorem 2. Let $\overline{V} = \{\overline{x}_1, \ldots, \overline{x}_n\}$ be any set of points in E^r such that

(i) o is the centroid of the points $\overline{x}_1, \ldots, \overline{x}_n$, and

(ii)   each open half-space of E^r bounded by a hyperplane through o contains at least two points of $\overline{V}$.

Then $\overline{V}$ is a Gale transform of the set of vertices of some $(n - r - 1)$-polytope.

Proof.   This depends upon the fact (which is also important in other applications) that the relationship between a set of points V and its Gale transform $\overline{V}$ is, in effect, symmetrical. This is, of course, obvious in the geometrical treatment of the last section, for $V = \text{vert } P$ and $\overline{V}$ are the images of vert T^{n-1} under orthogonal projections onto complementary perpendicular subspaces.

Let $\overline{V}$ be given satisfying the conditions of the theorem, and let $\underline{A}$ be the $r \times n$ matrix whose columns are $\overline{x}_1, \ldots, \overline{x}_n$. If y is an $n \times 1$ column vector, then the system of equations

$$\underline{A}y = o$$

has $n - r - 1$ affinely independent solutions, say $y_1, \ldots, y_{n-r-1}$. (By (i) there is an additional solution $(1, 1, \ldots, 1)'$.) If these solutions are written as the columns of an $n \times (n - r - 1)$ matrix $\underline{B}$, and $V = \{x_1, \ldots, x_n\}$ are the rows of $\underline{B}$, it follows that $\overline{V}$ is a Gale transform of V. Finally we note that, because of (ii) and Theorem 1, each point of V is a vertex of conv V; that is, V is the set of vertices of a convex polytope.

Further properties of Gale transforms are given in the next two theorems.

Theorem 3.   Let P be a polytope, and let $\overline{V}$ be a Gale transform of $V = \text{vert } P$. Then P is a pyramid with apex x if and only if $\bar{x} = 0$. Further, if P is a pyramid with basis Q, then $\bar{V}\setminus\{\bar{x}\}$ is a Gale transform of vert Q.

Proof. If P is a pyramid with apex x and basis Q, then $x \notin$ aff Q. Thus in any affine dependence (1) of the points of V, the coefficient of x must be zero. In other words, every such affine dependence must be essentially an affine dependence of the set vert Q. Thus the matrix $\underline{A}$(vert P) is formed from $\underline{A}$(vert Q) by adding a column of zeros corresponding to the point x. Since the arguments are reversible, this implies the statement of the theorem.

This result shows that a polytope P is an r-fold pyramid if a Gale transform of vert P has at least r points coinciding with o. (If P is not also an (r + 1)-fold pyramid, then exactly r points of $\bar{V}$ will coincide with o.) Theorems 1 and 3 enable us to describe completely the faces of a pyramidal polytope in a very simple manner, confirming the analysis of §2.3 (ii).

Theorem 4. Let P be a polytope, F a face of P, and Z the corresponding coface, and let $\bar{V}$ be a Gale transform of V = vert P. Let L be the complementary linear subspace in aff $\bar{V}$ perpendicular to aff $\bar{Z}$, and let Π denote orthogonal projection onto L. Then $(\bar{V}\setminus\bar{Z})\Pi$ is a Gale transform of vert F.

Proof. Let F be a j-face of the d-polytope P, let $V = \{x_1, \ldots, x_n\}$ as before, and suppose, without loss of generality, that vert $F = \{x_{r+1}, \ldots, x_n\}$. Then the set of affine dependences of vert F form an $(n - r) - j - 1$ dimensional vector space D', which may be identified with a subspace of D(V) (the space of affine dependences of V). Hence we may choose the basis $\{a_1, \ldots, a_{n-d-1}\}$ of D(V), with

$$a_k = (0, \ldots, 0, \alpha_{k,r+1}, \ldots, \alpha_{kn}),$$

for $k = 1, \ldots, n - r - j - 1$. In other words, the matrix $\underline{A}(V)$ takes the form

$$\underline{A}(V) = \left(\begin{array}{c:c} 0 & \underline{A}(\text{vert } F) \\ \hdashline B & C \end{array} \right) \begin{array}{l} \} n - r - j - 1 \\ \} j + r - d \end{array}$$

$$\underbrace{}_{r} \underbrace{}_{n - r}$$

Writing $z = (\zeta_1, \ldots, \zeta_{n-d-1})$ for a point of aff $\overline{V}$, we see that

$$\text{aff } \overline{Z} = \{ z \in \text{aff } \overline{V} \mid \zeta_1 = \ldots = \zeta_{n-r-j-1} = 0 \},$$

and that the colums of the $(n - d - 1) \times (n - r)$ matrix

$$\left(\begin{array}{c} \underline{A} \text{ (vert } F) \\ \hdashline 0 \end{array} \right) \begin{array}{l} \} n - r - j - 1 \\ \} j + r - d \end{array}$$

are the images of $\overline{x}_{r+1}, \ldots, \overline{x}_n$ under orthogonal projection Π onto the coordinate subspace

$$(\text{aff } \overline{Z})^\perp = \{ z \in \text{aff } \overline{V} \mid \zeta_{n-r-j} = \ldots = \zeta_{n-d-1} = 0 \}.$$

This proves the theorem.

The above proof clearly applies to a Gale transform of any set of points, and so by a slight modification we obtain the following result, which is useful in that it enables us, for any set of points X, to deduce the Gale transform of the set vert P of vertices of the polytope P = conv X.

Theorem 5. <u>Let P be a polytope in E^d, let $x \in P$, and let $\overline{X}$ be a Gale transform of the set vert P $\cup$ {x}. Let H be the hyperplane in aff $\overline{X}$ through o with normal $\overline{x}$, and let Π be the orthogonal projection onto H. Then $(\overline{X} \setminus \{\overline{x}\})\Pi$ is a Gale transform of vert P.</u>

It follows that, given a Gale transform $\overline{X}$ of an arbitrary point set X, by repeated orthogonal projections onto hyperplanes orthogonal to those points $\overline{x} \in \overline{X}$ such that

$$o \notin \text{relint conv}(\overline{X} \setminus \{\overline{x}\})$$

(so that there is some open half-space containing $\overline{x}$ alone), we eventually obtain a Gale transform of vert (conv X).

3.3 GALE DIAGRAMS

For many purposes Gale transforms are not sufficiently general, since we often require them to have certain convenient properties. We therefore generalize Gale transforms in the following way.

Definition. Two sets of points $\overline{V} = \{\overline{x}_1, \ldots, \overline{x}_n\}$ and $\overline{V}' = \{\overline{x}'_1, \ldots, \overline{x}'_n\}$ in E^{n-d-1} such that $o \in \text{int conv } \overline{V}$, $o \in \text{int conv } \overline{V}'$, are said to be <u>isomorphic, with</u> $\overline{x}_i$ <u>corresponding</u>

to $\bar{x}'_i$ (i = 1, ..., n), if for each pair of corresponding subsets $\bar{X} \subseteq \bar{V}$, $\bar{X}' \subseteq \bar{V}'$, the relations

$$o \in \text{relint conv } \bar{X}, \quad o \in \text{relint conv } \bar{X}'$$

either both hold or neither holds.

For example, from Theorems 1 and 1A, we see that Gale transforms of the set of vertices of a polytope P defined geometrically and algebraically are isomorphic sets in the above sense. (In fact it can be shown that a Gale transform in the geometric sense is also one in the algebraic sense. For details of this, see [6].) This motivates the following definition:

Definition. A <u>Gale diagram</u> of a d-polytope P with n vertices is any set of n points in E^{n-d-1} which is isomorphic to a Gale transform of vert P.

Gale diagrams are more general than Gale transforms, for, as we have seen, in the latter the centroid of the points is the origin o. In particular, if $\bar{V} = \{\bar{x}_1, \ldots, \bar{x}_n\}$ is a given Gale transform of vert P, and $\mu_1, \ldots, \mu_n$ are positive scalars, then it is clear that $\bar{V}' = \{\mu_1 \bar{x}_1, \ldots, \mu_n \bar{x}_n\}$ is isomorphic to $\bar{V}$, and so is a Gale diagram of P. In general, of course, the centroid of the points of $\bar{V}'$ will not be o, and therefore $\bar{V}'$ will not be a Gale transform. However, it is not difficult to see that the points of any Gale diagram can be multiplied by suitable positive scalars to convert it into a Gale transform. From the definition we immediately have the fundamental result:

Theorem 6. *Two polytopes P_1 and P_2 are combinatorially equivalent if and only if Gale diagrams $\overline{V}_1$ of P_1 and $\overline{V}_2$ of P_2 are isomorphic.*

Most of the results of §3.1 and §3.2 apply equally to Gale diagrams, the proofs either following directly from the definitions, or from the application of the condition of Theorems 1 and 1A for a subset of a Gale transform to correspond to a coface. We summarize these for reference:

Theorem 7. (i) *If $\overline{V}$ is a Gale diagram of P, then $Z \subset \text{vert } P$ is a coface of P if and only if*

$$o \in \text{relint conv } \overline{Z} .$$

(ii) *A set $\overline{V}$ of n points in E^{n-d-1} is a Gale diagram of a d-polytope P with n vertices if and only if every open half-space of E^{n-d-1} bounded by a hyperplane through o contains at least two points of $\overline{V}$.*

(iii) *If F is a facet of P, and Z is the corresponding coface, then in any Gale diagram of P, $\overline{Z}$ is the set of vertices of a (non-degenerate) simplex with o in its relative interior.*

(iv) *A polytope P is simplicial if and only if, for every hyperplane H containing o,*

$$o \notin \text{relint conv } (\overline{V} \cap H) .$$

(v) *A polytope P is an r-fold pyramid if and only if, in any Gale diagram of P, r points coincide with the origin o.*

Our next theorem is a generalization of Theorem 3. Let P be a polytope, let $V = \text{vert } P = \{x_1, \ldots, x_n\}$, and suppose, without loss of generality, that $x_n = o$, the origin. Let H be a hyperplane that strictly separates o from $\{x_1, \ldots, x_{n-1}\}$, and for $i = 1, \ldots, n-1$, let $y_i = H \cap \text{conv } \{o, x_i\}$. Writing $Y = \{y_1, \ldots, y_{n-1}\}$, then conv Y is a vertex-figure $H \cap P$ of P at the vertex x_n. Define the scalars $\mu_1, \ldots, \mu_{n-1}$ by

$$y_i = \mu_i x_i, \quad i = 1, \ldots, n-1,$$

so that $0 < \mu_i < 1$, and let

$$\begin{aligned} \lambda_1 y_1 + \ldots + \lambda_{n-1} y_{n-1} &= o, \\ \lambda_1 + \ldots + \lambda_{n-1} &= 0 \end{aligned}$$

be any affine dependence of Y. Then

$$\lambda_1 \mu_1 x_1 + \ldots + \lambda_{n-1} \mu_{n-1} x_{n-1} + \left(-\sum_{i=1}^{n-1} \lambda_i \mu_i\right) o = o$$

is an affine dependence of V. From the algebraic definition of the Gale transform, it then follows that we obtain a Gale transform $\bar{Y}$ of Y from a Gale transform $\bar{V}$ of V by omitting $\bar{x}_n$ and multiplying each other $\bar{x}_i \in \bar{V}$ by some positive scalar. Thus to find a Gale diagram of the vertex figure $H \cap P$, we remove $\bar{x}_n$ from a Gale diagram of P, and apply the process described in Theorem 5. Using this argument, and Theorem 4, and recalling the inductive proof of Theorem 16 of Chapter 2, we obtain the following comprehensive result:

Theorem 8. Let P be a polytope, $F_1 \subset F_2$ faces of P, $Z_1 = \text{vert } F_1$, and Z_2 the coface of P corresponding to F_2. In any Gale diagram $\overline{V}$ of P, let L be the orthogonal complementary (linear) subspace to aff $\overline{Z}_2$ in aff $\overline{V}$, and let Π be the orthogonal projection onto L. Then $(\overline{V} \setminus (\overline{Z}_1 \cup \overline{Z}_2))\Pi$ is isomorphic to a Gale transform of a set of points whose convex hull is F_2/F_1. To find a Gale diagram of F_2/F_1, we apply the process described in Theorem 5.

In future investigations we shall find it convenient to use a particular type of Gale diagram, which we shall call 'standard'. (This nomenclature differs from that of Grünbaum, who uses the term 'Gale diagram' for what we call 'standard Gale diagram'.) As we have seen, if we multiply the points of any Gale diagram of a polytope P (and, in particular, those of a Gale transform of vert P) by non-zero scalars, we obtain another Gale diagram of P. Consequently, there is a Gale diagram isomorphic to any given diagram, consisting of a subset of the points of $S^{n-d-2} \cup \{o\}$, where $S^{n-d-2} = S(o, 1)$ is the unit sphere in E^{n-d-1} with centre o. In fact, given the Gale diagram $\overline{V} = \{\overline{x}_1, \ldots, \overline{x}_n\}$ of the d-polytope P, we put

$$\hat{x}_i = o \text{ if } \overline{x}_i = o,$$

$$\hat{x}_i = \overline{x}_i / \|\overline{x}_i\| \text{ if } x_i \neq o.$$

The diagram so obtained is a standard Gale diagram of P.

In the case of d-polytopes P with $d + 2$ vertices, each Gale diagram is 1-dimensional, and a standard Gale diagram of P (in this case unique apart from sign) consists of $r + 1$ points and s + 1 points at unit distance on either side of o, and t points at o, where r, s, t are numbers such that r ≥ 1, s ≥ 1, t ≥ 0 and r + s + t = d (see Figure 20).

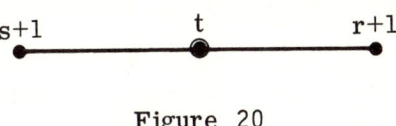

Figure 20

If t > 0, then P is a t-fold d-pyramid whose basis is the (r + s)-polytope whose standard Gale diagram is shown in Figure 21.

$$\underset{s+1}{\bullet}\underline{\hspace{2cm}}\underset{}{\circ}\underline{\hspace{2cm}}\underset{r+1}{\bullet} \quad (r \geq 1, \ s \geq 1)$$

Figure 21

This polytope is the convex hull of two simplices T^r and T^s, lying in complementary subspaces of E^{r+s}, such that $T^r \cap T^s$ is a single relatively interior point of each simplex. This polytope will be denoted by $T^{r,s}$.

The above is a complete description of all combinatorial types of d-polytope with d + 2 vertices, and leads immediately to their enumeration. For, we need only determine the number of integral solutions of the inequalities

$$r \geq 1, \quad s \geq 1, \quad r + s \leq d.$$

Recalling that, by Theorem 7, simplicial polytopes correspond to the case t = 0 (that is, r + s = d), we obtain:

Theorem 9. There are $[\frac{1}{4} d^2]$ distinct combinatorial types of d-polytopes with d + 2 vertices. Of these, $[\frac{1}{2}d]$ are the simplicial polytopes $T^{r,s}$ (r + s - d, r ≥ 1, s ≥ 1) described above, and the remainder are t-fold pyramids (t > 0) with simplicial bases $T^{r,s}$ (r + s = d - t, r ≥ 1, s ≥ 1).

In the case of d-polytopes with d + 3 vertices, the situation is a little more complicated. A standard Gale diagram of such a polytope P appears, for example, as in Figure 22, with p, q, ..., v integers representing the multiplicities of the points.

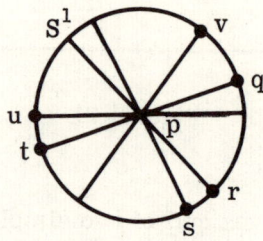

Figure 22

By Corollary 1 of Theorem 1, at least two points of the Gale diagram $\overline{V}$ must lie on each side of every diameter of the circle S^1, and by Theorem 7, if points lie at both ends of the same diameter, or at the centre o, then the polytope P represented by the diagram is not simplicial.

There are a number of operations that we can perform on a standard Gale diagram $\overline{V}$ which transform it into an isomorphic diagram:

(i) We may alter the angles between the diameters of S^1 containing points of $\overline{V}$ as long as we do not alter their order. For convenience of representation, we shall usually suppose that the angles between every diameter and its neighbours are equal.

(ii) If two adjacent diameters have points at one end only (these ends not being separated by a diameter containing points of $\overline{V}$), so that S^1 has at least four diameters containing points of $\overline{V}$, then these two diameters may be moved into coincidence. For example, the diameters labelled r and s in Figure 22 may be moved into coincidence, the resulting diameter having an end point with multiplicity r + s, as in Figure 23.

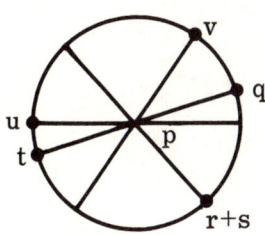

Figure 23

Conversely, any diameter with, say, r points at one end only can be replaced by r distinct diameters with a single end point. By these two processes we may minimize or maximize the number of diameters in the standard Gale diagram, and we obtain, respectively, the contracted Gale diagram and distended Gale diagram.

Notice however that we cannot perform this operation on a diameter marked at both ends, nor can we move into coincidence or transpose adjacent diameters containing points only at opposite ends, without changing the combinatorial type of the corresponding polytope. The last operation will prove important in the next section.

Bearing in mind Theorem 7, and the operations just described in (i) and (ii), we see immediately that each simplicial d-polytope with d + 3 vertices is represented by an essentially unique contracted Gale diagram of one of the forms in the sequence illustrated in Figure 24.

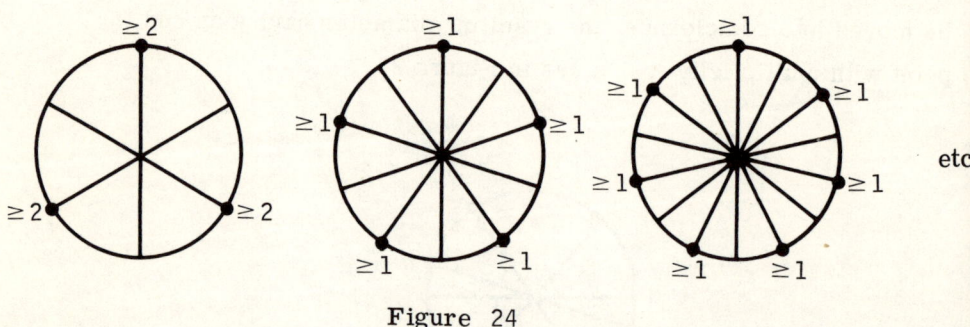

etc

Figure 24

The diagrams in the sequence have an odd number of diameters, and the points lie on alternate diameters around the circle. In this way we obtain a complete description of all the combinatorial types of simplicial d-polytopes with d + 3 vertices. In 1966 M. A. Perles considered the corresponding distended Gale diagrams, and proved the following theorem:

Theorem 10. The number $c_s(d + 3, d)$ <u>of distinct combinatorial types of simplicial d-polytopes with d + 3 vertices is given by</u>

$$c_s(d + 3, d) = 2^{[\frac{1}{2}d]} - [\tfrac{1}{2}(d + 1)] + \frac{1}{4(d+3)} \sum \phi(h) 2^{(d+3)/h} ,$$

where the summation is over all the odd divisors h of d + 3, and $\phi(h)$ is Euler's function (equal to the number of positive integers less than, and prime to h).

The problem of enumerating the non-simplicial d-polytopes with d + 3 vertices, where the corresponding standard Gale diagrams are allowed to carry points at both ends of a diameter, or at the centre, is more difficult, and no exact numerical result is known.* However, Perles has found an asymptotic formula for this number; the formula is quoted in 'Convex polytopes', §6.3.

The problem of enumerating combinatorial types of d-polytopes with more than d + 3 vertices seems completely intractable. The only results in this direction are for small values of d (see Tables 1 and 2 of 'Convex polytopes'), and for polytopes with certain symmetries (see [7]).

3.4 POLYTOPES WITH d + 2 AND d + 3 VERTICES

We are now in a position to prove two special cases of the Upper Bound Conjecture, namely that the cyclic polytopes C(d + 2, d) and C(d + 3, d) have at least as many k-faces ($1 \le k \le d - 1$) as any other polytope with the same dimension and the same number of vertices.

As we saw in §2.4, the number of k-faces ($1 \le k \le d - 1$) of C(v, d) ($v \ge d + 1$) is the same as the number of k-faces of any simplicial neighbourly (that is, $[\frac{1}{2}d]$-neighbourly) d-polytope with the same number of vertices. If P is an n-neighbourly

---

* Added in Proof: An expression for this number has recently been found by E. K. Lloyd: 'The number of d-polytopes with d + 3 vertices', Mathematika 17 (1970), 120-132.

d-polytope with v vertices, then any v - n vertices of P form
a coface. In other words, in any Gale diagram $\overline{V}$ of P, the origin
lies in the relative interior of the convex hull of any v - n points
of $\overline{V}$. Now any v - d - 2 points of $\overline{V}$ (which is a set of points in
E^{v-d-1}) are contained in some hyperplane H through o, and the
previous condition then implies that at least n + 1 points of the
remaining d + 2 must lie on each side of H (for otherwise we
would have some subset of v - n points lying entirely in one of
the closed half-spaces bounded by H, contradicting the fact that
the corresponding v - n vertices of P form a coface, which
corresponds to a simplicial (n - 1)-face of P). It is clear that
this condition is also sufficient for P to be n-neighbourly, for
then given any subset $\overline{X}$ of v - n points of $\overline{V}$, every hyperplane
through o has points of $\overline{X}$ on both sides, and so

o $\in$ relint conv $\overline{X}$.

We begin then by considering d-polytopes with d + 2
vertices. Since we are concerned with maximizing the number of
faces, by Theorem 25 of Chapter 2 we may restrict our attention
to simplicial polytopes. As we saw in §3.3, a standard Gale diagram
of such a polytope consists of (say) the point +1 taken r + 1 times
and the point -1 taken s + 1 times, where r + s = d,
r $\geq$ 1, s $\geq$ 1 (see Figure 21). It will be convenient to refer to
this diagram by the symbol (r, s), and, as before, to the
corresponding polytope by $T^{r,s}$. A k-face of $T^{r,s}$ (which is a
k-simplex) has k + 1 vertices, and so the transform of the
corresponding coface has d - k + 1 points, at least one of which is
+1, and at least one of which is -1. From this we can calculate
the number of k-faces of $T^{r,s}$ (r + s = d):

$$f_k(T^{r,s}) = \sum_{\substack{t \geq 1, u \geq 1 \\ t+u=d-k+1}} \binom{r+1}{t}\binom{s+1}{u}$$

$$= \binom{d+2}{d-k+1} - \binom{r+1}{d-k+1} - \binom{s+1}{d-k+1} . \quad (1)$$

Let us compare the number of faces of $T^{r,s-1}$ with those of $T^{r-1,s}$. From (1) we have

$$f_k(T^{r-1,s}) - f_k(T^{r,s-1}) = \binom{r}{d-k} - \binom{s}{d-k} .$$

If $r \geq s$ this number is non-negative, and if $r > s$ and $k > d - r$ (so certainly for $k = d - 1$, since we must have had $r \geq 2$) it is strictly positive. It follows immediately that a simplicial d-polytope with $d + 2$ vertices that has the largest number of k-faces ($k = 1, \ldots, d - 1$) corresponds to the standard Gale diagram (r, s) (where $r + s = d$) for which r and s are as nearly equal as possible, that is to say, when $r = [\frac{1}{2}(d + 1)]$ and $s = [\frac{1}{2}d]$. By the remarks at the beginning of the section, this already shows that the corresponding polytope $T^{r,s}$ is $[\frac{1}{2}d]$-neighbourly, and since the standard Gale diagram (r, s) is unique, and the cyclic polytope $C(d + 2, d)$ is also $[\frac{1}{2}d]$-neighbourly, this shows that $T^{r,s} \approx C(d + 2, d)$.

This can also be seen directly as follows. We label the $r + 1$ points $+1$ with $\bar{x}_1, \bar{x}_3, \ldots$, and the $s + 1$ points -1 with $\bar{x}_2, \bar{x}_4, \ldots$, and the corresponding vertices of $T^{r,s}$ with $x_1, \ldots, x_{d+2}$. By Corollary 2 of Theorem 1, the transform of a coface corresponding to a facet of $T^{r,s}$ consists of one of the points $+1$ and one of the point -1, so the coface consists of $\{x_i, x_j\}$ (say), where i and j are of opposite parity. But if we write $x_i = x(t_i)$ ($i = 1, \ldots, d + 2$), then this is seen to be precisely Gale's evenness condition (Proposition 18 of Chapter 2), and hence $T^{r,s} \approx C(d + 2, d)$.

Summarizing, we see that we have proved:

Theorem 11. Of all the combinatorial types of d-polytopes with d + 2 vertices, the cyclic polytopes C(d + 2, d) have the largest possible number of k-faces for each k = 1, ..., d - 1.

We now consider the more difficult problem of d-polytopes with d + 3 vertices, with a view to maximizing the number of their k-faces (k = 1, ..., d - 1). By Theorem 25 of Chapter 2 again, we need only consider simplicial polytopes, and their distended Gale diagrams have d + 3 diameters with one point on each. As we saw in §3.3, we may transpose a pair of adjacent diameters of the Gale diagram to give an isomorphic diagram, as long as the points of the diagram on the diameters lie at the same end. If they lie at opposite ends, then in general we change the combinatorial type of the corresponding polytope. We investigate the change in the number of faces involved in such a transposition.

Suppose then that $\overline{V}$ is a 2-dimensional distended Gale diagram with d + 3 points on distinct diameters. Let L and M be adjacent diameters of $\overline{V}$, and let $\overline{x} \in L$, $\overline{y} \in M$ be the points of $\overline{V}$ at opposite ends of these diameters (see Figure 25).

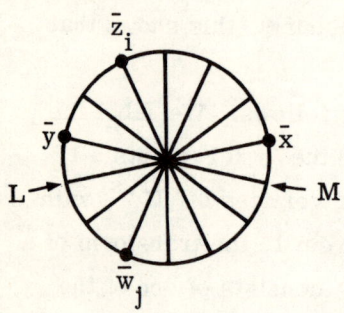
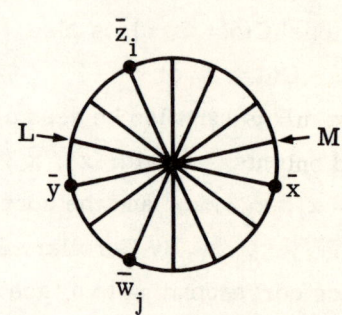

Figure 25   Figure 26

The remaining points of $\bar{V}$ fall into two sets, according to which side of L and M they lie. There are (say) r points $\bar{z}_1, \ldots, \bar{z}_r$ such that

$$o \notin \text{int conv } \{\bar{x}, \bar{y}, \bar{z}_i\}, \quad (i = 1, \ldots, r), \tag{2}$$

and s points $\bar{w}_1, \ldots, \bar{w}_s$ such that

$$o \in \text{int conv } \{\bar{x}, \bar{y}, \bar{w}_j\}, \quad (j = 1, \ldots, s), \tag{3}$$

so that $r + s = d + 1$. Notice that since $\bar{V}$ is a Gale diagram of some polytope P, we must have $r \geq 1$ and $s \geq 2$. If x, y, $z_1, \ldots, z_r, w_1, \ldots, w_s$ are the corresponding vertices of P, then for $i = 1, \ldots, r$, $\{x, y, z_i\}$ is not a coface of P, and for $j = 1, \ldots, s$, $\{x, y, w_j\}$ is a coface.

We now see what changes we make in P when we transpose the diameters L and M, as in Figure 26. For this to give us a Gale diagram of another polytope, we must have $r \geq 2$ also. It is obvious that the cofaces of P which contain at most one of x and y (but not both) are unaltered. However, the situations in (2) and (3) are now reversed, so that if u is any vertex of P ($u \neq x$, $u \neq y$), then $\{x, y, u\}$ is a coface of the new polytope Q if and only if it was not a coface of P.

The change in the number of k-faces of P ($k = 1, \ldots, d - 1$) is found as follows. A k-face of P containing neither x nor y (corresponding to a coface of P containing x and y) corresponds to a subset $\bar{X} \subseteq \bar{V}$ of $d - k + 2$ points containing $\bar{x}$ and $\bar{y}$, such that

$$o \in \text{int conv } \bar{X}. \tag{4}$$

147

If the subset $\overline{X}$ contains t of the points $\{\overline{z}_1, \ldots, \overline{z}_r\}$, and u of the points $\{\overline{w}_1, \ldots, \overline{w}_s\}$, then $t \geq 0$ and, by (4), $u \geq 1$. Hence the number of such subsets is

$$\sum_{\substack{t \geq 0,\, u \geq 1 \\ t+u=d-k}} \binom{r}{t}\binom{s}{u} . \tag{5}$$

After the transposition of L and M, we must similarly choose $t \geq 1$ points from $\{\overline{z}_1, \ldots, \overline{z}_r\}$, and $u \geq 0$ from $\{\overline{w}_1, \ldots, \overline{w}_s\}$ with $t + u = d - k$. The number is thus

$$\sum_{\substack{t \geq 1,\, u \geq 0 \\ t+u=d-k}} \binom{r}{t}\binom{s}{u} , \tag{6}$$

and from (5) and (6), and the remarks in the previous paragraph, we conclude that the increase in the number of k-faces is

$$\binom{r}{d-k} - \binom{s}{d-k} .$$

This number is clearly non-negative if $r \geq s$, and strictly positive if $r > s$ and $k \geq d - r$ (and so, since $r \geq 2$, certainly for $k = d - 1$).

We now investigate the circumstances under which we can increase the numbers of faces by such transpositions. We number the diameters in order $L_1, \ldots, L_{d+3}$. Suppose that, for $i = 1, \ldots, d + 3$, we have m_i points of $\overline{V}$ on one side of L_i, and n_i points on the other (so that $m_i + n_i = d + 2$). We write

$$e(\overline{V}) = \sum_{i=1}^{d+3} |m_i - n_i| .$$

148

Clearly for each i, $|m_i - n_i|$ and d have the same parity, so that the minimal value of $e(\overline{V})$ is 0 when d is even, and d + 3 when d is odd. By the remarks at the beginning of the section, it is clear that, in these cases, $\overline{V}$ is a Gale diagram of a simplicial neighbourly polytope. Diagrams in which the minima are obtained can be constructed as follows. If d is even, then the points of $\overline{V}$ are placed on the d + 3 diameters, so that points on adjacent diameters lie at opposite ends, as in Figure 27, which illustrates the case d = 4.

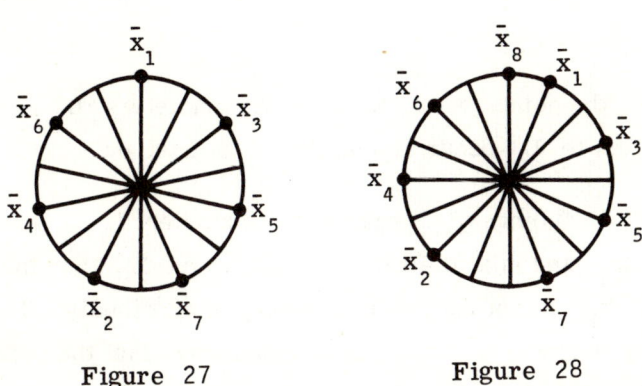

Figure 27   Figure 28

It is clear, moreover, that this arrangement is the only one that achieves the minimum, for if two adjacent diameters are marked at the same end, then at least one of the corresponding values of $|m_i - n_i|$ is greater than 2. In the case when d is odd, we omit one diameter, then mark the remainder as in the case d - 1. The remaining diameter is then marked at either end, as in Figure 28, which illustrates the case d = 5. In fact these diagrams correspond to the cyclic polytope C(d + 3, d); the diameters are labelled in order $L_1, \ldots, L_{d+3}$, so that in the case when d is odd, the two adjacent diameters with points at the same end are labelled L_1 and L_{d+3}. The correspondence $x_i \leftrightarrow x(t_i)$ between the vertices of the corresponding polytopes and the vertices of $C(d + 3, d)$ is then seen, by Gale's evenness condition, to give rise to a combinatorial equivalence.

We now suppose that $\overline{V}$ is a Gale diagram of a simplicial d-polytope P with $d + 3$ vertices, such that $e(\overline{V})$ does not take its minimal value (0 if d is even, $d + 3$ if d is odd). Then for some $i = 1, \ldots, d + 3$, $|m_i - n_i| \geq 2$ is maximal, and we may suppose $m_i > n_i$. It is clear that if either of the following occur

(i) the points $\overline{x}_{i-1}, \overline{x}_{i+1}$ on the diameters L_{i-1}, L_{i+1} adjacent to L_i occur among the n_i points, or

(ii) the point $\overline{x}_{i+1}$ (say) on the diameter L_{i+1} occurs among the m_i, and is at the same end as $\overline{x}_i$ on L_i, then the value $|m_i - n_i|$ is not maximal, contrary to hypothesis. Therefore we may suppose, without loss of generality, that the point $\overline{x}_{i+1}$ of $\overline{V}$ on the diameter L_{i+1} adjacent to L_i occurs among the $m_i > n_i$ points, and that $\overline{x}_i$ and $\overline{x}_{i+1}$ are at opposite ends of their respective diameters, as in Figure 29.

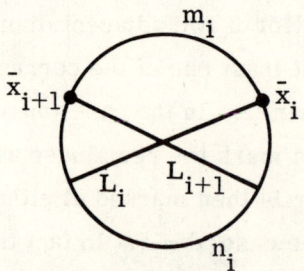

Figure 29

Then clearly we also have $|m_{i+1} - n_{i+1}| = |m_i - n_i| \geq 2$. If we now transpose the diameters L_i and L_{i+1} as above, the corresponding values of r and s are $m_i - 1$ and n_i respectively, so that $r > s$. The transposition alters no value $|m_j - n_j|$ except when $j = i$ or $i + 1$, and these are each decreased by 2. Consequently for the new diagram $\overline{V}'$, we have

$$e(\overline{V}') = e(\overline{V}) - 4,$$

while the number of k-faces of the corresponding polytope is strictly increased for $k \geq d - m_i + 2 \geq d - 1$.

Summarizing, the process of transposition applied in this way simultaneously decreases $e(\overline{V})$ to its minimum, and increases the number of k-faces of P (for at least $k = d - 1$). Since, as we remarked before, the diagrams $\overline{V}$ for which the minimal value of $e(\overline{V})$ is attained correspond to the simplicial neighbourly d-polytopes with $d + 3$ vertices, which have the same number of faces of each dimension as the cyclic polytopes $C(d + 3, d)$, we conclude:

Theorem 12. Of all the combinatorial types of d-polytope with $d + 3$ vertices, the cyclic polytopes $C(d + 3, d)$ have the largest possible number of k-faces for each $k = 1, \ldots, d - 1$.

4. The Upper Bound Conjecture for Spherical Complexes

4.1 SPHERICAL COMPLEXES

We begin by recalling the statement of the Upper Bound Conjecture. As in §2.3 for brevity we shall write $f_k(v, d)$ instead of $f_k(C(v, d))$ for the number of k-faces of a cyclic d-polytope with v vertices (or of any simplicial $[\frac{1}{2}d]$-neighbourly d-polytope with v vertices). Writing $\mathscr{P}^d$ for the set of all d-polytopes we define

$$\mu_k(v, d) = \max \{f_k(P) | P \in \mathscr{P}^d \text{ and } f_0(P) = v\},$$

the largest number of k-faces of a d-polytope with v vertices.

In 1957, in an abstract published in the Bulletin of the American Mathematical Society, T. S. Motzkin made the following conjecture ([8]).

The Upper Bound Conjecture (U.B.C.).

$$\mu_k(v, d) = f_k(v, d)$$

<u>for all</u> k, d, v <u>satisfying</u> $1 \le k < d < v$.

In §3.4 (Theorem 12) we showed that the U.B.C. was true if $v \le d + 3$. The purpose of this chapter is to show how, by using combinatorial arguments depending heavily on the Dehn-

Sommerville equations, a number of other cases can also be proved.

Before stating these in detail, it seems worth while to point out that all the results of this chapter (unlike those of §3.4 which depend on Gale diagrams, and those of Chapter 5 which depend on the shelling of polytopes) are valid in a wider context; they hold for spherical complexes as well as for polytopes.

By a <u>spherical polytope</u> in the unit $(d-1)$-sphere S^{d-1}, we mean the intersection of any finite set of closed hemispheres which is not empty and contains no pair of antipodal points of S^{d-1}. The <u>dimension</u> of a spherical polytope, and its <u>faces,</u> are defined in the obvious manner. By a <u>spherical complex</u> $\mathscr{C}$ we mean a finite set $\mathscr{C} = \{C_1, \ldots, C_r\}$ of distinct spherical polytopes <u>(cells)</u> C_i on S^{d-1}, which satisfies the following two conditions

(i) $\bigcup_{i=1}^{r} C_i = S^{d-1}$

(ii) For each i, j, the intersection $C_i \cap C_j$ is a face (proper or improper) of both C_i and C_j.

To each d-polytope P corresponds a spherical complex. For let $o \in \text{int } P$, and S^{d-1} be the unit $(d-1)$-sphere centred at o. Consider the radial projection $\Phi : \text{bd } P \to S^{d-1}$ defined by

$$x\Phi = R(o, x) \cap S^{d-1}$$

($R(o, x)$ is the ray passing through x, with end-point o). Then Φ maps the proper faces of P into spherical polytopes and the set of images of all the proper faces of P forms a spherical complex.

However all spherical complexes cannot be constructed by radial projection of polytopes in this way, even in a combinatorial sense. (An example of such a complex may be constructed from examples of [1, §11.5].) On the other hand, if the spherical complex is simplicial (that is, each cell is a spherical simplex), then the numbers of cells of each dimension are related by equations formally identical with the Dehn-Sommerville equations, and this is true whether the complex is the radial projection of a polytope or not. It will be noted that, in Theorem 1 below, our arguments depend only on computations from these equations, and so are true for all spherical complexes.

From §2.5 we recall that in discussing the problem of maximizing the number of faces of a polytope, it is sufficient to restrict attention to simplicial polytopes (and, in a similar way, for spherical complexes, we need only consider simplicial ones). Consequently, if we write $\mathscr{P}_s^d$ for the set of all simplicial d-polytopes, we see that

$$\mu_k(v, d) = \max \{f_k(P) | P \in \mathscr{P}_s^d \text{ and } f_0(P) = v\}.$$

4.2   PARTICULAR CASES OF THE U.B.C.

The following theorem gives the extent of present knowledge in the case of spherical complexes (though the arguments are stated in the language of polytopes). For the proof of the U.B.C. for polytopes, see Chapter 5.

Theorem 1.   <u>The U.B.C. holds in at least the following cases.</u>

(i)   <u>for every k, $1 \leq k < [\frac{1}{2}d]$.</u>

(ii)   <u>for every $k = n + p$, and</u>

(a) $d = 2n$ <u>and</u> $v \geq n - p - 2 + \frac{p+1}{p+2} n(n+1)$,

(b) $d = 2n + 1$ <u>and</u> $v \geq n - p - 2 + \frac{p+1}{p+2} (n+1)(n+2)$.

(iii) <u>for each</u> k, $1 \leq k < d$ <u>and</u> $v \leq d + 3$.

(iv) <u>for each</u> k, $1 \leq k < d$ <u>and</u> $d \leq 8$.

(v) <u>for</u> $k = d - 1$ <u>and</u>

(a) $d = 2n$ <u>and</u> $v \geq n^2 - 2$

(b) $d = 2n + 1$ <u>and</u> $v \geq (n+1)^2 - 3$.

(vi) (a) <u>for</u> $d = 2n$, $k = n$ <u>and</u> $v \geq \frac{1}{2}(n^2 + 3n - 6)$,

(b) <u>for</u> $d = 2n + 1$, $k = n$ <u>and</u> $v \geq \frac{1}{2}(n^2 + 5n - 4)$.

(vii) <u>for</u> $k = [\frac{1}{2}d]$, $v = d + 4$, $d + 5$ <u>or</u> $d + 6$, <u>if</u> $d \leq 76$.

(viii) <u>for</u> $k = [\frac{1}{2}d]$ <u>and</u> $d = 9, 10, 11$ <u>or</u> 12.

We shall begin by examining the particular cases $d = 3, 4$ and 5, which follow easily from the Dehn-Sommerville equations. These will be used without comment. Where no confusion can arise, we shall write f_k for $f_k(P)$.

(1) <u>$d = 3$.</u> For $P \in \mathcal{P}_s^3$, with $f_0 = v$,
$f_1 = 3v - 6$,
$f_2 = 2v - 4$.
Thus for each $P \in \mathcal{P}^3$, with $f_0 = v$,
$f_1 \leq 3v - 6 = f_1(v, 3)$,
$f_2 \leq 2v - 4 = f_2(v, 3)$.
(The values of $f_k(v, d)$ used here and elsewhere follow directly from the results of §2.3 (vi).)

(2) <u>d = 4.</u> For $P \in \mathscr{P}_s^4$, with $f_0 = v$,

$$f_2 = 2f_1 - 2v \leq 2\binom{v}{2} - 2v,$$
$$f_3 = f_1 - v \leq \binom{v}{2} - v,$$

with equality, in each case, for a 2-neighbourly polytope. Hence for $P \in \mathscr{P}^4$, with $f_0 = v$,

$$f_1 \leq \tfrac{1}{2}v(v-1) = f_1(v, 4),$$
$$f_2 \leq v(v-3) = f_2(v, 4),$$
$$f_3 \leq \tfrac{1}{2}v(v-3) = f_3(v, 4).$$

Equality in any relation characterizes the polytope as 2-neighbourly, and implies equality in the other relations.

(3) <u>d = 5.</u> For $P \in \mathscr{P}_s^5$, with $f_0 = v$,

$$f_2 = 4f_1 - 10v + 20,$$
$$f_3 = 5f_1 - 15v + 30,$$
$$f_4 = 2f_1 - 6v + 12,$$

and $f_1 \leq \binom{v}{2}$, with equality only for a 2-neighbourly polytope. Hence for $P \in \mathscr{P}^5$,

$$f_1 \leq \tfrac{1}{2}v(v-1) = f_1(v, 5),$$
$$f_2 \leq 2v(v-6) + 20 = f_2(v, 5),$$
$$f_3 \leq \tfrac{5}{2}v(v-7) + 30 = f_3(v, 5),$$
$$f_4 \leq v(v-7) + 12 = f_4(v, 5).$$

Equality in any relation (except the first) characterizes the polytope as simplicial and 2-neighbourly, and implies equality in the other relations.

This simple approach, using the solutions of the Dehn-Sommerville equations, breaks down in the case $d = 6$, due to the fact that the terms involving f_1 occur with negative signs. Hence inserting $f_1 \leq \binom{v}{2}$ and $f_2 \leq \binom{v}{3}$ does not lead to the required inequalities for f_3, f_4 and f_5. However, we can overcome this difficulty by using the following lemma, which enables us to reduce any even-dimension case $d = 2n$ to the case $d = 2n - 1$.

Lemma 1. <u>For all</u> $0 \leq k < d < v$,

$$\mu_k(v, d) \leq \frac{v}{k+1} \mu_{k-1}(v - 1, d - 1).$$

Proof. Let $P \in \mathscr{P}_s^d$, with $f_0(P) = v$, and vert $P = \{F_1^0, \ldots, F_v^0\}$. Then for each i, the vertex figure P/F_i^0 is simplicial, with at most $v - 1$ vertices (exactly $v - 1$ for each i if and only if P is 2-neighbourly). As each k-face of P has $k + 1$ vertices we deduce

$$f_k(P) = \frac{1}{k+1} \sum_{i=1}^{v} f_{k-1}(P/F_i^0), \quad k = 0, \ldots, d - 1,$$

and by definition,

$$f_{k-1}(P/F_i^0) \leq \mu_{k-1}(f_0(P/F_i^0), d - 1)$$

$$\leq \mu_{k-1}(v - 1, d - 1),$$

from which the result follows.

If $d \geq 4$ is even, then from Proposition 19 of §2.3 (vi), writing $d = 2n + 2$, $k - 1$ for k and $v - 1$ for v, we obtain

$$f_{k-1}(v - 1, d - 1) = \frac{k + 1}{v} f_k(v, d),$$

for all $0 < k < d < v$. Hence, combining this equation with the statement of the lemma, and $\mu_{k-1}(v - 1, d - 1) = f_{k-1}(v - 1, d - 1)$, we obtain

$$f_k(v, d) \leq \mu_k(v, d) \leq \frac{v}{k+1} \mu_{k-1}(v - 1, d - 1) =$$

$$= \frac{v}{k+1} f_{k-1}(v - 1, d - 1) = f_k(v, d).$$

In other words, if the U.B.C. holds for $(v - 1, d - 1)$, then it also holds for (v, d), if $d \geq 4$ is even. In particular, since the U.B.C. holds for $d = 5$ (for each $v - 1$), it is also true for $d = 6$.

We now proceed to the cases $d \geq 7$. We need the following lemma.

Lemma 2. For every simplicial polytope P with v vertices, and all $0 \leq k \leq r$,

$$\binom{r + 1}{k} f_r \leq \binom{v + k - 1 - r}{k} f_{r-k}$$

(independent of the dimension of P). Further, equality holds if and only if P is $(r + 1)$-neighbourly.

Proof. We calculate, in two different ways, the number $g_{r, r-k}$ of incidences of an r-face and an (r-k)-face of P (that is, the number of pairs (F^r, F^{r-k}) for which $F^{r-k} \subseteq F^r$). Each r-face is an r-simplex, so that

$$g_{r,r-k} = \binom{r+1}{k} f_r. \tag{1}$$

On the other hand, by Theorem 10 of §2.1, if $F^{r-k} \subseteq F^r$ there is a $(k-1)$-face F^{k-1} of F^r such that

$$F^{r-k} \cap F^{k-1} = \emptyset, \quad F^r = P \cap \text{aff}(F^{r-k} \cup F^{k-1}).$$

Hence $g_{r,r-k}$ cannot exceed the number of $(k-1)$-faces of P disjoint from F^{r-k}; that is, putting $w = v - (r - k + 1)$ (w is the number of vertices of P not in F^{r-k}), it cannot exceed $\binom{w}{k}$. Thus

$$g_{r,r-k} \leq \binom{w}{k} f_{r-k}, \tag{2}$$

with equality if and only if every $(r-k)$-face and every $(k-1)$-face of P determine an r-face of P; that is, if and only if P is $(r+1)$-neighbourly. Inequalities (1) and (2) lead immediately to the statements of the lemma.

We can now prove Theorem 1.

Proof of Theorem 1. (i) is immediately clear, since no d-polytope can have more k-faces $(1 \leq k < [\tfrac{1}{2}d])$ than a simplicial neighbourly d-polytope with the same number of vertices.

For (ii) we follow the treatment given in [5]. Let us first consider the case $d = 2n + 1$. The solutions of the Dehn-Sommerville equations are

$$f_{n+p} = \sum_{q=0}^{n} (-1)^q \frac{n+p+2}{n+q+1} \chi(n, p, q) f_{n-q-1}, \quad p = 0,\ldots, n,$$
(3)

where

$$\chi(n, p, q) = \sum_{s \geq 0} \binom{n-s}{p} \binom{n-s+q+1}{n+1}.$$
(4)

By Lemma 2 (with $r = n - q - 1$, $k = 1$) for $q \geq 0$ we have

$$f_{n-q-2} \geq \frac{n-q}{v-n+q+1} f_{n-q-1},$$
(5)

with equality characterizing P as neighbourly. Substituting (5) in (3) for each even q, we have

$$f_{n+p} \leq (n+p+2) \sum_{q \text{ even}} \left\{ \frac{\chi(n, p, q)}{p+q+1} - \frac{n-q}{v-n+q+1} \frac{\chi(n, p, q+1)}{p+q+2} \right\} f_{n-q-1},$$
(6)

and if v is sufficiently large all the coefficients will be non-negative, and we can substitute $f_{n-q-1} \leq \binom{v}{n-q}$ to obtain the U.B.C. Equality will, in fact, characterize the polytope as simplicial and neighbourly. So, we investigate the sign of

$$\frac{\chi(n, p, q)}{p+q+1} - \frac{n-q}{v-n+q+1} \frac{\chi(n, p, q+1)}{p+q+2}.$$
(7)

We write (7) as

$$\frac{\binom{n}{p}\binom{n+q+1}{n+1}}{p+q+1} - \frac{n-q}{v-n+q+1} \left\{ \frac{\binom{n}{p}\binom{n+q+2}{n+1} + \binom{n-1}{p}\binom{n+q+1}{n+1}}{p+q+2} \right\}$$

$$+ \sum_{s \geq 1} \binom{n-s+q+1}{n+1} \left\{ \frac{\binom{n-s}{p}}{p+q+1} - \frac{n-q}{v-n+q+1} \frac{\binom{n-s-1}{p}}{p+q+2} \right\}.$$

So, the term (7) is non-negative provided that each of the following terms is non-negative:

$$\frac{v - n + q + 1}{p + q + 1} - \frac{n - q}{p + q + 2} \left(\frac{n + q + 2}{q + 1} - \frac{n - p}{n} \right), \qquad (8)$$

$$\frac{v - n + q + 1}{p + q + 1} - \frac{n - q}{p + q + 2} \cdot \frac{n - s - p}{n - s}, \quad s \geq 1. \qquad (9)$$

In fact, if (8) is non-negative, so is the term (9) for $s = 0$. Now (9) increases with s, for

$$\frac{n - s - p}{n - s} = 1 - \frac{p}{n - s}$$

decreases as s increases. Hence it is enough to consider the sign when $s = 0$, and so it is sufficient to consider (8). Now the sign of (8) is that of

$$v - n + q + 1 - (n - q) \frac{p + q + 1}{p + q + 2} \left(\frac{n + q + 2}{q + 1} + \frac{n - p}{n} \right)$$

$$= v - n + q + 1 - \frac{n - q}{n} \cdot \frac{p + q + 1}{q + 1} \cdot \frac{(2n-p)q + n(n+3) - p}{p + q + 2}$$

$$= v - n + q + 1 - \frac{n-q}{n} (1 + \frac{p}{q+1})(2n - p + \frac{n(n+3) - (2n-p-1)(p+2) + 2}{p + q + 2}) :$$

161

But $(2n - p - 1)(p + 2)$ takes its maximum value when $p = n - \frac{3}{2}$, and so

$$h = n(n + 3) - (2n - p - 1)(p + 2) + 2$$
$$\geq n(n + 3) - (n + \tfrac{1}{2})^2 + 2 > 0.$$

Hence as q increases, each of the terms

$$\frac{n-q}{n}, \quad \frac{p}{q+1}, \quad \frac{h}{p+q+2}$$

decreases, so that the whole term (8) increases. Thus its minimal value is taken when $q = 0$, and so (8) is always non-negative provided that

$$v - n + 1 - \frac{p+1}{p+2}(n(n+2) + n - p) \geq 0,$$

that is, if

$$v \geq n - 1 + \frac{p+1}{p+2}(n^2 + 3n - p)$$
$$= n - p - 2 + \frac{p+1}{p+2}(n+1)(n+2), \tag{10}$$

and this leads immediately to the assertion of the theorem.

The even-dimensional case follows directly from this using Lemma 1. However, the direct method analogous to that above is very similar. If $d = 2n$, the solutions of the Dehn-Sommerville equations are

$$f_{n+p} = \sum_{q=0}^{n-1} (-1)^q \frac{n-q}{p+q+1} \chi(n-1, p, q) f_{n-q-1}, \quad p = 0, \ldots, n-1,$$

and proceeding as before with the substitution (5), we see that the U. B. C. holds proved v is so large that

$$\frac{n-q}{p+q+1} \chi(n-1, p, q) - \frac{n-q}{v-n+q+1} \cdot \frac{n-q-1}{p+q+2} \chi(n-1, p, q+1) \quad (11)$$

is non-negative for each $q \geq 0$. However, apart from the non-zero multiple $n - q$, (11) is just the same as (7) for the case $n - 1$, $v - 1$. We conclude that the U. B. C. holds if

$$v \geq n - p - 2 + \frac{p+1}{p+2} n(n+1), \quad (12)$$

as asserted in the statement of the theorem.

It should be remarked that in the case $q = 0$, the term by term comparison reduces to considering the term (8) alone. It follows that the inequalities (10) and (12) are, in fact, the best possible by this method.

The proof of part (iii) of Theorem 1 for polytopes has already been given in §3.4, Theorems 11 and 12. The truth of (iii) for spherical complexes follows immediately from the observation that every $(d - 1)$-dimensional spherical complex with at most $d + 3$ vertices is combinatorially equivalent to the boundary complex of a d-polytope [16].

For part (iv), we have already dealt with the cases $d \leq 6$, and so by virtue of Lemma 1 it is sufficient to consider the case $d = 7$. For this we require a result of Kruskal.

Let r and k be positive integers. The k-<u>canonical representation</u> of r is the representation of r in the form

$$r = \binom{r_0}{k} + \binom{r_1}{k-1} + \ldots + \binom{r_i}{k-i} \quad (13)$$

where r_p is defined provided

$$r > \binom{r_0}{k} + \binom{r_1}{k-1} + \ldots + \binom{r_{p-1}}{k-p+1}$$

and

$$r_p = \max \left\{ s \,\middle|\, r \geq \binom{r_0}{k} + \ldots + \binom{r_{p-1}}{k-p+1} + \binom{s}{k-p} \right\}. \tag{14}$$

For given positive numbers r, k and j, we define

$$r^{\{j|k\}} = \binom{r_0}{j} + \ldots + \binom{r_i}{j-i}, \tag{15}$$

where $r_0, \ldots, r_i$ are the numbers defined by the k-canonical representation (13) of r.

Let $\mathscr{C}$ be a simplicial complex. We define

$$\left.\begin{aligned}\kappa(r; k; j) &= \max \{f_j(\mathscr{C}) | f_k(\mathscr{C}) = r\} \text{ if } j > k, \\ &= \min \{f_j(\mathscr{C}) | f_k(\mathscr{C}) = r\} \text{ if } j < k.\end{aligned}\right\} \tag{16}$$

Then Kruskal's result, which is purely combinatorial, states

Lemma 3. <u>For all non-negative numbers r, k, j, the relation</u>

$$\kappa(r; k; j) = r^{\{j+1|k+1\}}$$

<u>holds.</u>

We shall not prove this result here; for a shorter proof then Kruskal's original, the reader is referred to G. Katona [3].

We now use Kruskal's Lemma 3 to prove that $\mu_k(v, 7) = f_k(v, 7)$. We consider the case $k = 6$; the Dehn-Sommerville equations show that

$$f_6 = 2(f_2 - 4f_1 + 10v - 20).$$

We shall show that $f_6 \leq f_6(v, 7)$.

For $v \leq 10$ this has been shown in part (3) of the theorem; hence we may assume $v \geq 11$. Since $f_6(11, 7) = 70$, we need only consider simplicial 7-polytopes with $f_6 \geq 70$. The 7-canonical representation of 70 is

$$70 = \binom{9}{7} + \binom{8}{6} + \binom{6}{5}.$$

So, by Lemma 3,

$$f_2 \geq 70^{\{3|7\}} = \binom{9}{3} + \binom{8}{2} + \binom{6}{1} = 118.$$

Hence, for those polytopes under consideration, the 3-canonical representation of f_2 will be of one of the types

$$f_2 = \binom{a}{3} + \binom{b}{2} + \binom{c}{1},$$

$$f_2 = \binom{a}{3} + \binom{b}{2},$$

$$f_2 = \binom{a}{3},$$

where $a \geq 9$. If it is of the third type, then by Lemma 3 again,

$$f_1 \geq \binom{a}{3}^{\{2|3\}} = \binom{a}{2},$$

$$v \geq \binom{a}{3}^{\{1|3\}} = a.$$

Since for $a \geq 9$ we have

$$\binom{a}{3} - 4\binom{a}{2} \leq \binom{a+1}{3} - 4\binom{a+1}{2},$$

it follows that

$$f_6 \leq 2\{\binom{a}{3} - 4\binom{a}{2} + 10v - 20\}$$

$$\leq 2\{\binom{v}{3} - 4\binom{v}{2} + 10v - 20\}$$

$$= f_6(v, 7),$$

and the U.B.C. follows.

Thus we may assume that the 3-canonical representation of f_2 is of the first or second type; that is

$$f_2 = \binom{a}{3} + \binom{b}{2} + \binom{c}{1},$$

with the term in c possibly absent, and with $a > b > 0$. Hence using Lemma 3,

$$v \geq f_2^{\{1|3\}} = \binom{a}{1} + \binom{b}{0} > a.$$

For fixed v, f_6 clearly increases or decreases with $f_2 - 4f_1$. Using Kruskal's Lemma again,

166

$$f_2 - 4f_1 \geq f_2 - 4f_2 \quad \{2|3\}$$

$$= \{\binom{a}{3} + \binom{b}{2} + \binom{c}{1}\} - 4\{\binom{a}{2} + \binom{b}{1} + \binom{c}{0}\},$$

with the term in c possibly absent. If c does occur, increasing it to its limit b - 1 clearly increases the right hand side of the expression above. Hence

$$f_2 - 4f_1 \leq \binom{a}{3} + \binom{b}{2} + \binom{b-1}{1} - 4\{\binom{a}{2} + \binom{b}{1} + 1\}$$

$$= \binom{a}{3} + \binom{b+1}{2} - 1 - 4\{\binom{a}{2} + \binom{b+1}{1}\}$$

$$< \binom{a}{3} + \binom{b+1}{2} - 4\{\binom{a}{2} + \binom{b+1}{1}\}.$$

So, we must find an upper bound for the last expression, under the assumption $b < a$. Since we keep a fixed, we therefore wish to maximize

$$\binom{b+1}{2} - 4\binom{b+1}{1} = \tfrac{1}{2}(b+1)(b-8).$$

This can clearly be done by taking b to be as large as possible, as long as we may take $b \geq 7$. Since b can take any value up to a - 1, and since $a \geq 9$, the maximum is attained for $b = a - 1$. Substituting this value gives

$$\binom{a}{3} + \binom{a}{2} - 4\{\binom{a}{2} + \binom{a}{1}\} = \binom{a+1}{3} - 4\binom{a+1}{2}.$$

Thus, since $a < v$, and the expression on the right increases for $a \geq 9$, we deduce

$$f_2 - 4f_1 < \binom{a+1}{3} - 4\binom{a+1}{2} \leq \binom{v}{3} - 4\binom{v}{2}.$$

In other words,

$$f_6 < 2\{\binom{v}{3} - 4\binom{v}{2} + 10v - 20\}$$

$$= f_6(v, 7).$$

By similar reasoning we can prove that $\mu_k(v, 7) = f_k(v, 7)$ for $k = 3, 4$ and 5, but it is simpler to observe that the Dehn-Sommerville equations can be solved in a different way, to give

$$2f_3 = 5f_6 + 10f_1 - 30v + 60,$$
$$2f_4 = 9f_6 + 4f_1 - 12v + 24,$$
$$2f_5 = 7f_6;$$

hence these cases follow directly from the case $k = 6$. Notice also that the case $k = 3$ is also covered by parts (ii) and (iii) of the theorem. This completes the proof of part (iv).

We shall not discuss the remaining parts of the theorem here, they all depend upon recent work of Grünbaum [2] that has been done since the appearance of 'Convex Polytopes' [1]. Parts (v) and (vi) are proved using the following inequality, which is more sophisticated than that of Lemma 2:

$$k(k + 1)f_k + (v + 1 - k)f_{k-2} \leq k(v + 1 - k)f_{k-1}$$

(provided $f_{k-1} > 0$). It is likely that this would lead to a similar improvement of one value of v in the remaining cases of part (ii). Parts (vii) and (viii) are proved using Kruskal's Lemma 3 again.

5. The Upper Bound Conjecture for Polytopes

The proof to be given in this chapter was first formulated in July 1970, and follows closely the original paper by P. McMullen [13]. It depends on a new technique developed since the first four chapters of these notes were written, namely that of shelling the boundary complex of a polytope. This idea, which will be discussed in §5.2, was originated by H. Bruggesser and P. Mani [11].

In addition to this, it is necessary to use a reformulation of the Dehn-Sommerville equations, see §5.1. This reformulation was stated recently in a paper by McMullen and Walkup [15] though, in the form in which it is to be used, it appears in the works of Sommerville (see [1, Theorem 9.2.2]). Of particular importance is the definition of the quantities $g_k^{(e)}(P)$, and, in fact, it was through considering the properties of these integers (instead of the $f_k(P)$) that the proof of the U.B.C. for polytopes was discovered.

5.1 REFORMULATION OF THE DEHN-SOMMERVILLE EQUATIONS

Let P be a simplicial d-polytope. As in §2.4, we write

$$f(P, t) = \sum_{j=-1}^{d-1} (-1)^{j+1} f_j(P) t^{j+1},$$

with the usual convention $f_{-1}(P) = 1$. We shall find it convenient subsequently to adopt the additional conventions $f_j(P) = 0$ if

$j < -1$ or $j \geq d$. (Contrast earlier work, when we put $f_d(P) = 1$.) The Dehn-Sommerville equations are then equivalent to the relation

$$f(P, 1 - t) = (-1)^d f(P, t). \tag{1}$$

Following [15], for any integer $e \geq d$, we define

$$g^{(e)}(P, t) = (1 - t)^e f(P, \tfrac{t}{t-1}). \tag{2}$$

It is clear that $g^{(e)}(P, t)$ is a polynomial in t of degree e. If we write

$$g^{(e)}(P, t) = \sum_{k=-1}^{e-1} g_k^{(e)}(P) t^{k+1}, \tag{3}$$

then it is an easy calculation to show that, for $k = -1, \ldots, e - 1$,

$$g_k^{(e)}(P) = \sum_{j=-1}^{k} (-1)^{k-j} \binom{e - j - 1}{e - k - 1} f_j(P). \tag{4}$$

From (1) and (2), we deduce that

$$(-1)^{e-d} t^e g^{(e)}(P, t^{-1}) = (-1)^{e-d} t^e (1 - t^{-1})^e f(P, \tfrac{t^{-1}}{t^{-1}-1})$$

$$= (-1)^{e-d}(t - 1)^e f(P, 1 - \tfrac{t}{t-1})$$

$$= (1 - t)^e f(P, \tfrac{t}{t-1}) = g^{(e)}(P, t),$$

so that

$$(-1)^{e-d} t^e g^{(e)}(P, t^{-1}) = g^{(e)}(P, t). \tag{5}$$

From (3) and (5), we deduce that for $k = -1, \ldots, [\tfrac{1}{2}e] - 1$,

$$g_k^{(e)}(P) = (-1)^{e-d} g_{e-k-2}^{(e)}(P) . \tag{6}$$

It may easily be verified that

$$f(P, t) = (1 - t)^e g^{(e)}(P, \tfrac{t}{t-1}) , \tag{7}$$

and so, for $j = -1, \ldots, d - 1$,

$$f_j(P) = \sum_{k=-1}^{j} \binom{e-k-1}{e-j-1} g_k^{(e)}(P) ; \tag{8}$$

thus the equations (6) are equivalent to the Dehn-Sommerville Equations.

In general, the equations (6) are not independent (in fact, we know from Theorem 20 of §2.4 that exactly $[\tfrac{1}{2}(d+1)]$ of them are). We shall be most interested in the case $e = d$, where the equations are independent, except that if $d = 2n$ is even, the $(n - 1)$-st equation is obviously redundent. (In [15], the case $e = d + 1$ is studied in some detail; we shall remark on the reasons in §5.3.)

We easily deduce from the definition (2) the recurrence relations

$$g_k^{(e+1)}(P) = g_k^{(e)}(P) - g_{k-1}^{(e)}(P) , \tag{9}$$

$$g_k^{(e)}(P) = \sum_{j=-1}^{k} g_k^{(e+1)}(P) . \tag{10}$$

Let us consider in more detail the case $e = d$. In view of the relation (6), we can rewrite the solution (8) in the form

$$f_j(P) = \sum_{k=-1}^{n-1} \{ \binom{d-k-1}{d-j-1} + (1 - \delta_{k, d-n-1})\binom{k+1}{d-j-1} \} g_k^{(d)}(P) ,$$
(11)

where $\delta_{k, d-n-1}$ is the Kronecker delta function. The coefficient of $g_k^{(d)}(P)$ in each relation is non-negative, and positive for each k if $j \geq n - 1$. (As usual, we write $n = [\frac{1}{2}d]$.) Now if P is neighbourly, then for $j = -1, \ldots, n - 1$,

$$f_j(P) = \binom{v}{j+1} ,$$
(12)

where $v = f_0(P)$, and so for $k = -1, \ldots, n - 1$,

$$g_k^{(d)}(P) = \sum_{j=-1}^{k} (-1)^{k-j}\binom{d-j-1}{d-k-1} \binom{v}{j+1}$$

$$= \binom{v-d+k}{k+1} .$$
(13)

(The details of this calculation are a little difficult. The easiest way of proving the result is to notice that $f(P, t)$ and $(1 - t)^v$ are polynomials which differ only in their terms of degree greater than n. The same is then true of $g^{(d)}(P, t)$ and

$$(1 - t)^d (1 - \tfrac{t}{t-1})^v = (1 - t)^{-(v-d)} ,$$

and the coefficient of t^{k+1} in the latter is just $\binom{v-d+k}{k+1}$.)

It thus follows that to prove the inequalities

$$f_j(P) \leq f_j(v, d) ,$$

172

it suffices to prove

(14) Lemma. Let P be a simplicial d-polytope with v vertices. Then, for $k = 1, \ldots, n-1$,

$$g_k^{(d)}(P) \leq \binom{v - d + k}{k + 1}.$$

Notice that equality for each k will imply that P is neighbourly.

5.2 SHELLING THE BOUNDARY COMPLEX

Let P be a d-polytope. We denote by ∂P the boundary complex of P, which is the geometric cell complex consisting of the proper faces of P, together with the empty set. (A cell complex is a collection of distinct polytopes in some euclidean space, with the properties that every face of a polytope in the complex belongs to the complex, and the intersection of any two polytopes in the complex is empty, or a face of each polytope.) A shelling of ∂P is a labelling of the facets of P, say $F_1, F_2, \ldots, F_m$ $(m = f_{d-1}(P))$, so that for $s = 2, \ldots, m-1$,

$$F_s \cap \bigcup_{t=1}^{s-1} F_t$$

is homeomorphic to a $(d-2)$-ball. This will imply that, for $s = 1, \ldots, m-1$, $\bigcup_{t=1}^{s} F_t$ is homeomorphic to a $(d-1)$-ball.

It has recently been proved ([11]) that ∂P is always shellable. We shall give an outline of the method of [11], referring the reader to that paper for details of the proof. Let L be any line meeting int P in general position with respect to the facets of P. That is, L meets the $m = f_{d-1}(P)$ hyperplanes of support corresponding to the facets of P in distinct points. Let z be a variable point on L. As z moves along L, starting at a point of int P, suppose that z meets successively the supporting hyperplanes $H_1, \ldots, H_m$ containing the facets $F_1, \ldots, F_m$. (Here we must regard order along L in the projective sense, so that z passes through infinity, and eventually returns to int P.) Then $F_1, \ldots, F_m$ is a shelling of P. In fact, $F_1, \ldots, F_s$ are just the facets of P 'visible' from a point z between H_s and H_{s+1} if z has not reached infinity, or the facets not 'visible' from z if z has passed through infinity. Further, $F_s \cup \bigcup_{t=1}^{s-1} F_t$ is that part of relbd F_s 'visible' (or not 'visible', as appropriate) in H_s from the point $L \cap H_s$.

Now let P again be a simplicial d-polytope, and let $F_1, \ldots, F_m$ be a shelling of ∂P. For $s = 1, \ldots, m$, let

$$M_s = \bigcup_{t=1}^{s} F_t ,$$

and let $f_j(M_s)$ denote the number of faces of P lying in M_s, with the convention $f_{-1}(M_s) = 1$. For $k = -1, \ldots, d-1$, let

$$g_k^{(d)}(M_s) = \sum_{j=-1}^{k} (-1)^{k-j} \binom{d-j-1}{d-k-1} f_j(M_s) . \tag{15}$$

We consider, for $s = 1, \ldots, m$, the quantity

$$g_k^{(d)}(M_s) - g_k^{(d)}(M_{s-1}) ,$$

where we shall write $g_k^{(d)}(M_0) = 0$ for all k. For $s = 2, \ldots, m-1$, $F_s \cap M_{s-1}$ is a topological (d - 2)-ball in the boundary of the (d - 1)-simplex F_s, and so is the union of certain facets of F_s.

The intersection of these facets is a face of F_s, of dimension $(d-r-2)$ (say), and the facets are just those which contain this face. In going from M_{s-1} to M_s, the faces of P in M_s which are not in M_{s-1} are just those which contain the opposite r-face F of F_s. The number of such j-faces is clearly

$$f_{j-r-1}(F_s/F) = \binom{d-r-1}{d-j-1},$$

since F_s/F is a $(d-r-2)$-simplex. Thus, for $s = 2, \ldots, m-1$, and $k = -1, \ldots, d-1$,

$$g_k^{(d)}(M_s) - g_k^{(d)}(M_{s-1}) = \sum_{j=-1}^{k} (-1)^{k-j} \binom{d-j-1}{d-k-1} \binom{d-r-1}{d-j-1}$$

$$= \sum_{j=-1}^{k} (-1)^{k-j} \binom{d-r-1}{d-k-1} \binom{k-r}{j-r}$$

$$= \delta_{k,r}, \tag{16}$$

the usual Kronecker delta. Inspection shows that this relation also holds in the extreme cases $s = 1$ (with $r = -1$) and $s = m$ (with $r = d-1$). That is, in going from M_{s-1} to M_s, we increase $g_r^{(d)}$ by one, and (for $k \neq r$) each other $g_k^{(d)}$ by zero.

Two immediate consequences of this argument are as follows. Firstly, for $k = -1, \ldots, d-1$,

$$g_k^{(d)}(P) = g_k^{(d)}(M_m) \geq 0.$$

Secondly, since the reverse labelling $F_m, \ldots, F_1$ of the facets of P may also easily be seen to be a shelling of ∂P, with the rôles of the r-face and $(d-r-2)$-face of F_s interchanged, we conclude that, for $k = -1, \ldots, n-1$,

$$g_k^{(d)}(P) = g_{d-k-2}^{(d)}(P) \ ;$$

that is, we have another proof of the Dehn-Sommerville equations.

If P has v vertices, then $g_0^{(d)}(P) = v - d$, and the crucial Lemma (14) is clearly an immediate consequence of

(17) Lemma. For $1 \le k \le d - 1$,

$$(k + 1) g_k^{(d)}(P) \le (v - d + k) g_{k-1}^{(d)}(P) \ .$$

If x is a vertex of P, let P_x denote the vertex-figure of P at x. We shall prove Lemma (17) by comparing the values of $g_k^{(d-1)}(P_x)$ ($x \in \text{vert } P$) and $g_k^{(d)}(P)$ in two ways.

Firstly, we obtain the exact relationship

$$\sum_{x \in \text{vert } P} g_{k-1}^{(d-1)}(P_x) = (k+1)g_k^{(d)}(P) + (d-k)g_{k-1}^{(d)}(P) \ . \qquad (18)$$

This can be proved algebraically by writing down the expression for $g_{k-1}^{(d-1)}(P_x)$, and using the relationship

$$\sum_{x \in \text{vert } P} f_{j-1}(P_x) = (j+1) f_j(P) \ .$$

We prefer to give a more geometrical proof. A shelling of ∂P will clearly give rise to a shelling of each ∂P_x. Let $F_1, \ldots, F_m$ be a shelling of ∂P, and suppose that (as described above), in adding F_s to M_{s-1} to obtain M_s, we increase $g_r^{(d)}$ by one (and no other $g_k^{(d)}$). Clearly d vertex-figures of P are affected. In $r + 1$ of them, we contribute one to $g_{r-1}^{(d-1)}$, since we are adding on the $(r - 1)$-face of the r-face F of F_s which does not contain the given vertex; in the remaining $d - r - 1$ of them, we contribute one to $g_r^{(d-1)}$. Considering the increase in $\sum g_{k-1}^{(d-1)}(P_x)$ leads at once to (18).

The second relationship is

$$\sum_{x \in \text{vert } P} g_{k-1}^{(d-1)}(P_x) \leq v\, g_{k-1}^{(d)}(P). \tag{19}$$

For, consider any vertex-figure P_x. We may clearly choose the shelling of ∂P by the method of [11] so that at some stage M_s consists of all the facets of P which contain x. (We make the line L pass through some point beyond all the facets of P which contain x, and beneath all the remaining facets of P.) It is now easy to see that, in the induced shelling of ∂P_x, a contribution to $g_{k-1}^{(d-1)}$ gives rise to a contribution to $g_{k-1}^{(d)}$ for ∂P. Thus

$$g_{k-1}^{(d-1)}(P_x) \leq g_{k-1}^{(d)}(P),$$

and summing over the vertices of P yields (19) immediately.

The expressions (18) and (19) at once give the statement of Lemma (17), and by virtue of the remarks before Lemmas (14) and (17), this proves the Upper Bound Conjecture.

5.3 REMARKS

The proofs of the various parts of the Upper Bound Conjecture given in Chapter 4 are valid for spherical complexes as well as for polytopes. However, the existence of non-shellable triangulations of the $(d-1)$-sphere (for $d \geq 4$) show that the proof given here does not apply to spherical complexes. In that respect, then, the Upper Bound Conjecture remains open.

The Upper Bound Conjecture is part of the far wider problem of determining all possible f-vectors $(f_0(P), \ldots, f_{d-1}(P))$ of d-polytopes P. As far as the restricted problem for simplicial polytopes is concerned, the numbers $g_k^{(d+1)}(P)$ seem to be more important, and in [15] is made the

(19) Generalized Lower Bound Conjecture. <u>Let P be a simplicial d-polytope. Then for $k = 0, \ldots, n-1$,</u>

$$g_k^{(d+1)}(P) \geq 0.$$

From (9), we see that an equivalent conjecture is that

$$g_k^{(d)}(P) \geq g_{k-1}^{(d)}(P).$$

The usual Lower Bound Conjecture for simplicial polytopes, discussed in [1, §10.2], is a consequence of this conjecture. Some evidence is offered for the conjecture in [15]; in particular, it is shown that no sharper linear inequalities are satisfied by the f-vectors of all simplicial d-polytopes.

Even more intriguing, if rather less plausible, is the conjecture proposed in [14]. Let r and k be positive integers, and, as in §4.1, let

$$r = \binom{r_0}{k} + \binom{r_1}{k-1} + \ldots + \binom{r_i}{k-i}$$

be the k-canonical representation of r. If j is also a positive integer, we define

$$r^{\langle j | k \rangle} = \binom{r_0 + j - k}{j} + \binom{r_1 + j - k}{j-1} + \ldots + \binom{r_i + j - k}{j-i}.$$

We also define $0^{\langle j|k\rangle} = 0$. If $(f_0, \ldots, f_{d-1})$ is a d-vector, we write for $k = -1, \ldots, d$,

$$g_k = \sum_{j=-1}^{k} (-1)^{k-j} \binom{d-j}{d-k} f_j ,$$

with the conventions $f_{-1} = 1$, $f_d = 0$. We then have

(20) Conjecture. $(f_0, \ldots, f_{d-1})$ <u>is the f-vector of a simplicial d-polytope if and only if</u>

$$g_k = -g_{d-k-1}, \quad k = -1, \ldots, [\tfrac{1}{2}(d-1)] ,$$

$$g_k \geq 0, \quad k = 0, \ldots, n-1 ,$$

$$g_k \leq g_{k-1}^{\langle k+1|k\rangle}, \quad k = 1, \ldots, n-1 .$$

By a method using Gale diagrams analogous to that of §3.4, the conjecture can be proved in case $f_0(P) \leq d + 3$ (see [14]).

References

For historical notes, and references to papers published before 1967, the reader should consult Grünbaum's book [1], which has an extensive bibliography.

[1] B. Grünbaum. Convex Polytopes. John Wiley and Sons, London-New York-Sydney, 1967.

[2] B. Grünbaum. 'Some results on the upper-bound conjecture for convex polytopes'. SIAM J. Appl. Math. 17 (1969), 1142-1149.

[3] G. Katona. 'A theorem of finite sets.' In Theory of Graphs, Proc. Colloq. Tihany, Sept. 1966, Ed. by P. Erdös and G. Katona, Akadémiai Kiadó, Budapest, 1968.

[4] I. G. MacDonald. 'Polynomials associated with finite cell complexes.' J. London Math. Soc. (to be published).

[5] P. McMullen. 'On the upper-bound conjecture for convex polytopes.' J. Combinatorial Theory (to be published).

[6] P. McMullen and G. C. Shephard. 'Diagrams for centrally symmetric polytopes.' Mathematika, 15 (1968), 123-138.

[7] P. McMullen and G. C. Shephard. 'Polytopes with an axis of symmetry.' Canad. J. Math. (to be published).

[8] T. S. Motzkin. 'Comonotone curves and polyhedra.' Abstract 111, Bull. Amer. Math. Soc. 63 (1957), 35.

[9] G. C. Shephard. 'A theorem on cyclic polytopes.' Israel J. Math. 6 (1968), 368-372.

[10] H. Tverberg. 'A generalization of Radon's Theorem.' J. London Math. Soc. 41 (1966), 123-128.

[11] H. Bruggesser and P. Mani. 'Shellable decompositions of cells and spheres,' (to be published).

[12] P. McMullen. 'On a problem of Klee concerning convex polytopes.' Israel J. Math. 8 (1970), 1-4.

[13] P. McMullen. 'The maximum numbers of faces of a convex polytope.' Mathematika 17 (1970), 179-184.

[14] P. McMullen. 'The numbers of faces of simplicial polytopes,' Israel J. Math. (to appear).

[15] P. McMullen and D. W. Walkup. 'A generalized lower-bound conjecture for simplicial polytopes,' (in preparation).

[16] P. Mani. 'On spheres with few vertices.' J. Combinatorial Theory (to appear).

Index

Affine basis 6
    dependence 2
    dimension 6
    hull 6
    transformation 14
Affinely span 6
Affinity 14

Ball 5
Beneath a hyperplane 113
Beyond a hyperplane 113
Bipyramid 77
Boundary complex 173
Bruggesser, H. (iii), 169

Carathéodory's Theorem 26
Central reflection 15
Centroid 10
Coface 121
Combinatorial equivalence 57
Complex, spherical 153
Cone, convex 7
Congruent transformation 15
Contracted Gale diagram 141
Convex cone 7
    dependence 2
    hull 8

    polytope 10
    set 3
Cube 80
Cyclic polytope 82

Dehn-Sommerville equations 101
Dependence, affine 2
    convex 2
    linear 1
    positive 6
Dimension 6, 8
Distended Gale diagram 141
Dual 61

Edge 39
Euler hyperplane 98
Euler's theorem 94
Evenness condition 85
Ewald, G. (iv)

f-vector 98
Face 39
Face, improper 39
Face-lattice 56
Face, proper 39
Facet 39

Few vertices, polytopes with
    143

Gale, D.   119
Gale diagram,   119, 135
    contracted   141
    distended   141
    standard   138
Gale transform   119, 120
Gale's evenness condition   85
Generalized lower bound
    conjecture   178
Grünbaum, B.   (iv), 168

Half-space,   29
    supporting   29
Helly's theorem   24
Hull, affine   6
    convex   8
    linear   6
    positive   7
Hyperplane,   7
    supporting   29

Improper face   39
Independence   2

Katona, G.   165
Kruskal, J. B.   163

Line   7
Linear dependence   1
Linear hull   6

Lower bound conjecture   178

MacDonald, I. G.   (iv), 103
Mani, P.   (iii), 169
Motzkin, T. S.   152

Nearest point map   31
Neighbourhood   5
Neighbourly polytopes   85, 90
Normal   7
Normal vector to half-space
    29

Orthogonal projection   18

Parallel   7, 15
Paralleloptope   80
Peano-Jordan volume   38
Perles, M. A.   119, 142, 143
Permissible projective trans-
    formation   20
Perpendicular   18
Point   7
Polar set   61
Polarity   61
Polyhedral set   43
Polytope, convex   10
    cyclic   82
    neighbourly   85, 90
    simple   82
    simplicial   81
    spherical   153

Positive hull 7
Positively span 7
Prism 78
Projective transformation 18
Projection, orthogonal 18
Proper face 39
Pulling the vertices of a
 polytope 116
Pyramid 75

Radon's theorem 22
Ray 31
Reflection, central 15

Shelling a polytope 173
Simple polytope 82
Simplex 10, 74
Simplicial polytope 81
Sphere 5
Spherical complex 153
 polytope 153
Standard Gale diagram 138
Subspace 3
Supporting half-space 29
Supporting hyperplane 29

Transformation, affine 14
 congruent 15
 projective 18
Translation 14
Tverberg, H. 24

Upper bound conjecture
 for polytopes 169
 for spherical complexes 152

Vertex 39
Vertex-figure 73
Volume of convex set 38

Walkup, D. W. 169

QA
691
M28

APR 26 1972